The Axemaker's Gift

THE AXEMAKER'S GIFT

Technology's Capture and Control of Our Minds and Culture

JAMES BURKE
ROBERT ORNSTEIN

Illustrations by Ted Dewan

A Jeremy P. Tarcher/Putnam Book
Published by G. P. Putnam's Sons
New York

To Madeline and Sally

Most Tarcher/Putnam books are available at special quantity discounts for bulk purchases for sales promotions, premiums, fund-raising, and educational needs. Special books or book excerpts also can be created to fit specific needs. For details, write or telephone Special Markets, The Putnam Publishing Group, 200 Madison Ave., New York, NY 10016; (212) 951-8891.

A Jeremy P. Tarcher/Putnam Book
Published by G. P. Putnam's Sons
Publishers Since 1838
200 Madison Avenue
New York, NY 10016
http://www.putnam.com/putnam

First Trade Paperback Edition 1997

Published simultaneously in Canada

Library of Congress Cataloging-in-Publication Data
Burke, James, date.
The axemaker's gift / James Burke, Robert Ornstein.
p. cm.
ISBN 0-87477-856-5
1. Technology—Social aspects. 2. Technology—History.
3. Culture—History. 4. Technology and civilization.
I. Ornstein, Robert E. (Robert Evan), date. II. Title.
HM221.B84 1995 95–2146 CIP
306.4'6—dc20

Book design by Debbie Glasserman
Cover design by Tom McKeveny

Cover illustration (top): Stone Age axehead from Acheul, France. Original photograph courtesy Corbis/Bettmann

Printed in the United States of America

10 9 8 7 6 5 4 3 2 1

This book is printed on acid-free paper. ♾

Acknowledgments

We would first like to thank Carolyn Doree for her many, many megabytes of research work for this book.

We have had the benefit of the work of many others, but especially the research of Alan Parker, John Wood, Jerome Burne (who also read the manuscript), and Lynne Levitan on specific topics.

The book was read and critiqued by a small army of readers, among whom we would like to especially thank Brent Danninger, Evan Neilsen, Howard Gardner, Bob Cialdini, Sally Mallam and Tom Malone. To those couple of dozen who wished to remain anonymous, your advice and suggestions were invaluable.

Ted Dewan did a magnificent job of translating our ideas into illustration, and Jane Isay encouraged, coddled, and critiqued the manuscript into a book.

C o n t e n t s

Prologue *xi*

I. Getting an Edge *1*
1. Getting an Edge *3*
2. Token Contribution *35*
3. The ABC of Logic *63*

II. Cutting up the World *89*
4. Faith of Power *91*
5. Fit to Print *121*
6. New Worlds *145*
7. Root and Branch *175*
8. Class Act *199*
9. Doctor's Order *225*

III. Picking up the Pieces *251*
10. Journey's End *253*
11. Forward to the Past *279*

Select Bibliography *313*

Index *339*

The Guides, the Wardens of our faculties
And Stewards of our labour, watchful men
And skillful in the usury of time,
Sages, who in their prescience would control
All accidents, and to the very road
Which they have fashion'd would confine us down
Like engines . . .

WILLIAM WORDSWORTH, *THE PRELUDE,* BOOK V

PROGRESS

PROLOGUE

In his Bostan, *Saadi of Shiraz stated an important truth when he told this miniature tale:*

A man met another, who was handsome, intelligent, and elegant. He asked him who he was. The other said: "I am the Devil."

"But you cannot be," said the first man, "for the Devil is evil and ugly."

"My friend," said Satan, "you have been listening to my detractors."

IDRIES SHAH, *REFLECTIONS*

This book is about the people who gave us the world in exchange for our minds. The gifts we accepted from them gave us the power to change the way we lived, but doing so also changed the way we thought. This Faustian bargain was sealed more than a million years ago, but as you will see, the bargain didn't turn out to be quite what either party might have expected.

We call those with whom we made the bargain "axemakers." But they make more than axes. They make everything. They make our hopes and dreams. They make what we love and what we hate. They make all this, because they make the tools that change our surroundings. And when their innovations are taken up and used, the effect is to shape the world in which we live, the beliefs for which we fight and die, the values we live by. And our very nature.

Originally the axemakers were ancient hominids who had the talent to reshape stones one piece at a time, and in doing so create tools that would chop up the world. This axemaking ability to do things in the proper order is one of the brain's many natural talents. In our ancient past, the all-powerful axemaker talent for performing the precise, sequential process that shaped axes would later give rise to the precise, sequential thought that would eventually generate language and logic and rules, which would formalize and discipline thinking itself. The newly dominant sequential talent of the mind used the "cut-and-control-it" capability to extract more knowledge from the world and then use that knowledge to cause even further change. Thanks to the axemakers' talents and their gifts, at any time things literally would never be the same again.

Look about you and you will find evidence of the axemakers'

presence in anything you see. The whole of the natural world has been altered by them. The domestication of animals, breeding, horticulture, agriculture, irrigation, architecture, and mining are only a few of the ways in which the axemakers have literally changed the face of the planet.

Axemakers influence practically every aspect of your daily life. Their gifts not only change the world of their time but also remain in use, to affect later periods. Any modern environment is a mixture of these world-altering changes, whose origins range back thousands of years into the past. The fact that you are able to read this book originates with the effects of the fifteenth-century printing press. The food you ate for breakfast today was delivered to the supermarkets thanks to the nineteenth-century combustion engine. The clothes you're wearing now began their existence on a prehistoric loom. You are alive, in all probability, thanks to one or another medical advance dating from some time in the last hundred years.

Your workplace probably includes thirteenth-century paper, sixteenth-century lathe-turned furniture, nineteenth-century plastics, fifteenth-century toilets, seventeenth-century electricity powering nineteenth-century telephones, and early twentieth-century computers. The water supply in the restrooms is delivered by sixteenth-century pumping systems. The paint on the office walls contains nineteenth-century artificial dyes. Your business itself likely runs on a top-down decision hierarchy dating from command structures first established seven thousand years ago to run the first city-states. Your fellow workers probably enjoy male-female relationships influenced by the first ancient stone tools.

By getting into our life, the axemakers get into our heads. Although secondary, the social effects of axemaker innovation also shape those aspects of our lives that cannot be so readily observed. In the fifteenth century, when the astronomer Copernicus rocked the boat with his view of a universe that did not have humankind at its center, he cut the ground from under the established authority of the church. In the late eighteenth century, when the Industrial Revolution suddenly brought hundreds of thousands of farm workers into city factories, their potentially disruptive presence provoked draconian legislation, which included capital punishment for such minor offenses as stealing a handkerchief. In nineteenth-century Western culture, where mechanization was taken to be the sign of an advanced society, those communities unable to adapt easily to technology, or for some other reason remote from it, were regarded as inferior in all aspects of life. Today, television coverage of the glitterati provides influential role models for behavior, while television soap operas offer glimpses of a world whose values many viewers admire and adopt.

Thanks to technological advance, we regard a world of manicured lawns and freshly painted houses as "better" than the huts, bare earth, and midden heaps among which our medieval ancestors lived. That we may automatically expect our civil rights to be observed springs directly from the fact that such rights are embodied in laws which are disseminated and practiced throughout our society thanks to the invention of print and telecommunications technology. And we tend to dismiss the opinion of old people in part because five hundred years ago Gutenberg's press downgraded their social position, when books began to replace the oral tradition as the main repository of society's experience and accumulated knowledge.

Through history, when the axemakers changed our world in these ways, we were in most cases willing and eager participants in the matter. Most of the time the axemakers' gift was irresistible. More often than not it was a cure for a disease, or a faster way to do something, or a means to facilitate what we wanted to do. So we came back for more, unmindful of the other, not easily visible, changes the gift might eventually bring. But we could never unmake history, and with each gift there was no choice but to adapt to the effects of the change.

This has been true for every generation of our ancestors since the process began, well over a million years ago. When we used the first axemaker tool to cut more food from nature than nature offered on her own, we changed our future. As a result, there were soon many more of us. And as our numbers grew, so did the power of those who could wield the axe most effectively. They became leaders. Most of the rest followed the axe.

At first, the effect of the new tools on the world was insignificant. Early human beings lived in small, widely scattered groups. With their hand axes and spears they often hunted and harvested the immediate surroundings until the area was empty of food, at which time they moved on. The Earth was so rich and so vast that for a long time the damage caused by this indiscriminate axe was not worth consideration. But by twelve thousand years ago, this had already changed. The number of people and tools had reached a critical mass, and our presence began to make itself more widely felt, as the first villages began to effect permanent changes on the countryside around them, with domestication of animals and the first moves toward irrigation and agriculture. Soon after, the first clay tablet sign-writing was developed and the rate of change accelerated.

With the acceleration also came major change in the nature of the relationship between us and our axemakers. When tools like the alphabet appeared, they encouraged new ways of thinking. The linear nature of the alphabet facilitated sequential, reductionist, logical forms of expression and thought. Its ease of use promoted literacy and with it the involvement of citizens in the governing process. Above all, the alphabet made it possible to ask questions that were not immediately essential to the well-being of the community. These questions, about matters such as the origin of the universe, the nature of life, or the sum of the internal angles of a triangle, generated increasingly esoteric axemaker vocabulary. They also changed the way we thought about the way we thought.

Increasingly, specialist knowledge became prized in part for the very fact that it was understood by so few. Axemaker logic and reductionism created a widening gulf between axemakers and the societies they affected that would not be bridged until the axemakers themselves provided tools with which to build the bridges.

Meanwhile, the reductionist, "cut-and-control" approach to the world has generated myriad scientific and technological specialisms which have produced a wealth of glittering technologies. These in turn have created a domino effect whose negative impact can be seen in any crowded Third World city street today. Whereas twelve thousand years ago there were five million of us, today the same number is born every two weeks. By the year 2100, the effect of this growth on the body of the planet could well bring the irreversible loss of half the species on Earth.

In the face of so much promise, why has this happened? Because the axemakers' gifts were, like their axes, double-edged. The gifts offered our leaders and institutions opportunities for self-fulfillment, self-aggrandizement, self-indulgence, self-satisfaction, and self-enrichment that were so immediately seductive to those who sought power that they denied or ignored the potential long-term effects. Today while millions in the Third World live, thanks to the effects of sanitation, medicine, and the increasing population, they also starve. Ironically, at the same time, the developed nations use their immense technological and scientific capabilities to pave almost half their cultivable land.

The double-edged effect of the axe is beginning to make itself felt globally. As a result of the continued push for the (entirely laudable) aims of increased standards of living, better health, and adequate diet and mass-produced consumer goods, one-third of the world's forests have disappeared, the population is exploding, the oceans are becoming depleted, and the atmosphere is severely polluted. Human shortsightedness, axemaker knowledge, and the destruction of the environment are inextricably linked.

In the ancient past, when we came out of Africa and began to hack our way across the planet, led by tribal chieftains whose axe-maker tools gave them the power to cut-and-control the world, we did not realize (or did not care) how close we already might be to journey's end, to the day when we would discover that the world's resources were no longer limitless. For tens of thousands of years, it continued to be accepted practice to use axemaker's gifts to take what we wanted from the world, without recompense. Today, as a result, we in the industrialized nations are healthier, wealthier, better fed, more informed, and more mobile than anybody ever.

The fact that progress has also brought a measure of devastation in its wake should not surprise us, since as we progressed, we destroyed. Only rarely, if ever, did we look back to examine the effect of our passage on the world, because our axemakers always led us forward, toward the horizon we never expected to reach. Thanks to the axe, the past was dead and the future was ours to take. It is only now, in the remains of journey's end, that we have begun to question why we find ourselves compromised.

But with this question also comes the realization that if we can become more conscious of how axemaker gifts always unleashed the kind of power that first changes the environment and then changes our way of thinking and our values, we will recognize that our survival now depends on harnessing the same technological power to save ourselves. We already have the capability to do so and the necessary tools. All we need to do is become acquainted with the process by which technology changes minds, and then apply it to ourselves. That is the purpose of this book.

The changes in the world that change our minds don't need to be random and beyond our control. What we might be able to do, given the information we now have (and which we will present in this book) about how the environment develops the mind, is to change our minds by changing the environment of our children, whether this is through new use of multimedia computers, or by a more direct experience of nature, or through new forms of socialization. We stand at a point where for the first time we can consciously take our development in our own hands and use it to generate talents that will suit the world of tomorrow.

There are axemakers who use similar talents everywhere today. In general, they are people who produce the technology that improves the quality of our lives. Indeed, if we are to remedy some of the more catastrophic ecological and social circumstances in which we find ourselves at present, it will be only through their tools and with their help that we will succeed. Axemakers are not alien beings, but the key problem is that what axemakers know and how they express it is not well understood by the majority of people. It is essential that we come to understand the work of these axemakers and the process through which their activities shape our lives, so that society in general can begin socially to assess their work and direct it toward more wisely chosen ends. Thanks to the axemakers, their work, previously directed by king or priest or CEO or bank or institution, may now come to be controlled by the larger community. But only if we learn how to use the axemakers' gifts to our general and individual advantage.

This book begins with the first of those gifts: the axe of ancient Africa.

I.

GETTING AN EDGE

Chapter 1

GETTING AN EDGE

Whence did the wond'rous mystic art arise
Of painting speech and speaking to the eye?
That we by tracing magic lines are taught
How both to colour and embody thought?

THOMAS ASTLE, *THE ORIGINS AND PROGRESS OF WRITING,* 1803

The axemakers appeared about four million years ago here, on the only planet in the solar system capable of sustaining them. Their planetary life-support system was provided by energy from the sun, moved around the planet (then and now) by a turbulent web of complex, interactive energy cycles. These range from massive, continent-wide atmospheric disturbances to microscopic bacterial activity in plant roots. The constant and sometimes violent interaction among the cycles is pervasive and continuous, and we will touch on only a few elements of it in the next few pages.

The sun triggers the primary cycle, when solar radiation hits the surface of the upper atmosphere with the energy of an atomic explosion every square mile. Some of the energy radiates away into space, but enough of it reaches Earth's surface to keep everybody and everything alive. And since the Earth spins, when this life-sustaining energy hits any point on the surface, it pulses to a maximum and falls to a minimum once each twenty-four hours.

Because the Earth orbits the Sun at a tilt, the energy pulse is three times more powerful at the equator than at the Pole. That energy difference drives the next cycle: atmospheric circulation.

As the air moves across the ocean, some of its energy shifts to the sea in the form of surface drift currents, or waves, that interact with the tides caused by lunar and solar cycles. All of these oceanic movements contribute to the sea-temperature cycle because the ocean behaves like the atmosphere, with cold water moving south at depth from the poles, to upwell at the equator and then return northwards close to the surface. Deep-water storms also occasionally blow up and scour vast areas of ocean floor, lifting and moving thousands of tons of sediment and marine life.

The ocean and the atmosphere propel the atmospheric gas cycles. The one that keeps us breathing is the oxygen cycle. Oxygen is stored in the atmosphere and the ocean and is generated by three different production cycles. The first happens in the upper atmosphere, where the solar energy pulse splits water molecules and releases their oxygen; the second is the product of twenty-four-hour-cycle plant photosynthesis; and the third is long-cycle, generated by dying organisms in the sea, releasing oxygen that enters the atmosphere across the sea surface-atmosphere boundary.

The atmosphere cycle triggers evaporation and precipitation between the ocean and the atmosphere, and that generates the vital freshwater cycle. Thirteen thousand cubic kilometers of fresh water are stored in the atmosphere as water vapor that rises and condenses around particles of dust. This starts cloud-building cycles that are driven by the amount of vapor, or the local temperature, or the air pressure, or the amount of heat in the cloud. When the vapor moves inland on the wind, it rises into cooler air and condenses as rain that eventually returns to the ocean (through evaporation from soil, vegetation, or lakes and streams, or from the ocean itself), either by seepage, infiltration to underground aquifers and springs, or through soil drainage and run-off.

Rain also triggers complex microcycles involving electrochemical reactions in rocks, weathering them, and releasing their minerals. Some of them are dissolved in run-off, some taken up by plant roots and later recycled to the soil in fallen leaves, some leached away by the freshwater cycle to underground aquifers.

In this kind of constantly changing environment, an organism only survives if it can take energy where it can get it. So the successful types evolve to take advantage of the form of food available where they happen to be. The others go the way of anything in nature that stands still or doesn't adapt: they die.

The clearest example of adaptation is the way some plants open and close in the morning and at night, but plants also interact with the environment in other more idiosyncratic ways. Stone plants in Namibia avoid being eaten by cattle because they look like pebbles;

mimosa become smaller and less visible when they're touched; some orchids look and feel like female insects, so male insects attempt to mate with them and take away the plant's pollen.

However, nature is no free lunch. At any level in the hierarchy of life, each time an organism taps into the energy supply, only one-tenth of the energy available at that level passes on to the next level below. From the total amount of energy first photosynthesized by green plants and algae, the amount diminishes as it goes through half a million plant types, thirty million kinds of invertebrates, a hundred million different insects, and more than fifty thousand vertebrates. For those micro-organisms unfortunate enough to find themselves at the end of the line, only one ten-thousandth of the original energy snatched from the sky by chlorophyll is available for consumption.

From the planet-wide atmospheric energy cycle all the way down to the microsystems at work around plant roots, this great passage of energy through the life chain generates constant subcycles. For instance, North American bacteria living on plant roots promote the growth of foliage, which is the principal food source for the white-tailed deer. When the deer eat the foliage, they drop nitrogen-rich waste, on which the bacteria feed. But as the deer population grows, it is preyed on by wolves, and if the wolves do too well, the deer numbers begin to dwindle and the wolves begin to starve. So perhaps they chase lesser prey, like sheep. Then, when sheep numbers dwindle, the wolves go back to the now-recovered deer population, whose increased droppings have in the meantime promoted the foliage growth that provides them with food.

So this cycle and countless others like it constantly, haphazardly, begin and conclude. The combination of the cycles generates myriad different forms of usable energy that maintain the tremendous variety of species on Earth. This diversity ensures the long-term durability of the entire ecosystem, because a mix-and-match system is better equipped to adapt in the face of all the random changes that occur naturally. Win some, lose some.

In primeval times, thanks to this endless exchange, life cycled along for billions of years, adapting slowly to climatic change or unfortunate events, such as catastrophic meteor strikes. And then nature's basic adaptability was challenged by change for which the system could not compensate and from which it would never completely recover, because it was a new *kind* of change. It was sequential and cumulative, not cyclic.

This is how it happened. Roughly thirteen million years ago, several centuries of drought caused the forests of East Africa to thin out. This quirk of weather set in train a series of events that would put the entire ecosystem in the power of one single species, which would then use the power to sever its links with nature and eventually bring the planet to the edge of ruin.

The drier climate forced many tree-dwelling primates out of their forest homes and challenged them to adapt to new ecological niches in the expanding savannahs. Those primates that stayed in the forest evolved into chimpanzees, gorillas and a recently discovered intermediate species. The ones that moved out would become us. Some of them would become axemakers.

Establishing exactly where and what we came from is difficult. It is not easy to find the evidence for something that happened millions of years ago. So scientific understanding of these ancestral matters changes constantly. One day in 1993, for example, a dramatic new find caused everybody to rethink the timetable of human events when anthropologist Gen Suwa came across a fossil tooth in north-central Ethiopia. It turned out to be part of the oldest discovered human ancestor and Suwa's team called it *Ramidus.*

Whoever he was, *Ramidus* lived about four and a half million years ago, was about four feet tall, and had both simian and human characteristics. We don't know yet whether or not he walked upright. Contrary to our general understanding before he was found, *Ramidus* may have lived in forest lands, because his remains were found among plentiful tree seeds, petrified wood, and fossil antelopes and squirrels. Astoundingly, he is intermediate in development and has been called a "missing link" between the fully upright walkers that existed a million years later and the apes of six million years earlier. These are slow processes.

The evidence is scanty so far, but if *Ramidus* was indeed a forest dweller who rose on his hind legs to pluck fruit from the trees, that simple fact will force evolutionary biologists to revise the entire explanation for the origins of bipedal walking. In any case, the transition to walking upright seems to have happened around four million years ago with *Ramidus,* or with another ancestor a few hundred thousand years later. However, the important thing is not precisely when it happened but that it happened at all.

A three-and-a-half-million-year-old footprint in East Africa, discovered by Mary Leakey, indicates that, by that time, our human ancestors had clearly diverged from the great apes. The footprint is of a creature unquestionably standing on two legs. The adjustment from walking on all fours to walking upright encouraged reliance on vision, and freed the front limbs for other work, like toolmaking and carrying. The weight of the body, previously supported by the front limbs, shifted to the legs and pelvis, which thickened to carry the weight of the upper body. This, in turn, refashioned childbirth, causing babies to be born immature.

At this point, we know these ancestors were living in country like the modern East African savannah and they were no longer in the trees. In their new habitat, natural selection favored the ones with the ability to walk upright in the high grass and bush because they were better equipped to see predators and food (so they survived), and they could probably survive the heat better (so they kept their cool). The importance of sophisticated toe control, previously essential for life in the trees, diminished in importance in favor of sensitivity and manipulability in the hands. And the fingers became more and more dexterous, more and more able to make fine manipulations, including fine cuts.

With these developments came asymmetric limbs. These are, for obvious reasons, not advantageous in a quadruped, bird or aquatic mammal, since stronger limbs on one side would make movements eventually circular and the animal would get nowhere. With hominid feet, rather than forelimbs responsible for movement, the two forelimbs were now free to develop independently, and with this came the ability to have different skills and strengths in the right and the left.

This capability would be behaviorally vital in the repertoire of these early hominids, because this handy asymmetry was accompanied by brainy asymmetry. By three million years ago, the left side of the brain of the tiny *Australopithecus* differed from the right, the slightly larger left side handling manipulative, tool-making abilities.

The hands were now more precise, capable of complex movements. The eyes could see into the distance as well as coordinate hand movements and this led to an increase in the informational capacity of the brain. Busy brains are big brains, and so by two and a half million years ago hominid brain size had doubled. Two-handedness, coupled with an enhanced ability to process information in the brain, took hominids to the next stage of evolution. The new type is called *Homo habilis,* the key actor in the story.

Habilis changed the course of history, because they were able to shape pebbles into flint tools, and these tools could quickly and advantageously help them to manipulate their environment. It was this ability of these first axemakers that would break the cycle that had bound us to nature and that over the following two million years would imperil all life on the planet.

The first primitive tools, simple cobbles made by fractures and used 2.6 million years ago for cutting and scraping, were found in what is now Ethiopia. Then pebble-axes gave *habilis* the cutting edge with tools that would not only bring change to the environment but also release the tool-users forever from the slow development of natural processes. Now tools could supplant biological evolution as the main source of change.

Axes made it possible to build shelters and construct primitive settlements, and they physically changed the world once and for all. This, in turn, changed hominid behavioral patterns because the tools also permitted *habilis* to go hunting. More important, they went hunting in groups, and this was to prove a meaningful thing to do. First of all, it changed the working day and then it changed the menu. Previously, foraging in the bushes for enough fruits and berries to feed a small community would consume a great deal of time, but now a group of tool-using hunters could bring home enough food from a single chase to support several families on meat for days.

This food sharing encouraged *habilis* to establish a stable home base and a more permanent society. That they had brains capable of doing this relates to their ability to hunt in groups. Consider what hunting requires: speed and accuracy, obviously, but even more important, the ability to plan, communicate, and cooperate. These communicative abilities helped *habilis* to get better organized, but they also set the scene for greater things because they laid down the mental matrix necessary for thought and reason, language and culture.

Over millennia, the new species evolved and spread through and out of Africa. By approximately two million years ago, a five-foot tall, heavy-set descendant of the original hominids, a type called *Homo erectus,* whose skeleton from the neck down resembled modern humans, was living in the cooler East African hill country, running down its prey, ranging further in search of food.

It had taken between six and nine million years for the prehuman brain to grow enough for some form of communal living to develop and for the invention and use of tools. But once these tools and systems emerged, they interacted with each other and spurred more rapid change in the world and, as a consequence, in the way we thought.

The earliest proper stone tools, dating from the *erectus* period, are found in Kenya and Tanzania and were used for cutting and pounding vegetables, butchering meat, and cracking bones to get at the marrow inside. They were also used for sharpening animal bones to dig for roots.

Late in this period, our ancestors developed double-edged hand axes. Perhaps the axe had by that time brought into existence the division of labor, allowing males, for the first time, to hunt as well as to scavenge for animals killed by large predators. Females probably also scavenged, but no doubt spent much of their time digging for roots, gathering vegetables, and caring for the young.

Over the following million years, axemaking became more and more sophisticated. Seven hundred thousand years ago our predecessors also possessed a type of hand-axe that is found throughout Africa, the Middle East, most of Europe and India, and parts of Southeast Asia. A huge find at Kilombe, in Kenya, suggests that by this time the axemakers had already developed techniques for mass production. They were using some form of template to make hand axes of the same length but of different breadth. This kind of work demanded more and more attention and memory on the part of anybody learning the skill, so the grunts and hand waving that went with the teaching must at some stage have proved less than adequate. This drove some of the teachers to make more sophisticated use of an already developed anatomical ability they happened to have. They were able to make mouth noises.

Tools also promoted the evolution of speech in another way, thanks to *erectus'* discovery of fire. Six hundred thousand years ago, when brain size had doubled again, toolmaking often involved the use of the lips, teeth, tongue, and even the airways, when blowing on tinder. The arrangement of the protohuman larynx and nasal-mouth cavities had meant that bends in the airways made mouth breathing necessary at times of strenuous activity. As fire made cooking possible, softer food meant that molar teeth gradually became smaller, and the shape of the mouth and larynx changed.

Thanks to the new trick of pounding and grinding food, large teeth, and their powerful accompanying jaw muscles and bony anchor points, were no longer necessary, so they got smaller. This lightening of the bones of the skull had the effect of giving more space for an expansion of the brain, and it may have been because of this that speech could develop. The tongue also became more flexible, and together with the other new physical characteristics the ability to make more subtly controllable vocal sounds was enhanced.

This had an effect on anatomy, because over and above the

changes to the larynx and tongue, vocalization required greater control of the diaphragm and ribs, and that in turn selected for the enlarged nerve canals found in the spine of modern humans. So with all these changes, for the first time, the brain of early hominids was able to generate complicated thought and simple sounds.

As the axemakers changed us and the world with their tools, they also radically changed the way we all perceived the world. The tools changed the physical shape of the human brain. Over millions of years, the process of evolution selected the fundamental brain structure we have had for thousands of years, adept at detecting the bits of the world most useful for survival and reproduction, at least in the type of environment present when the brain evolved. For that reason, we notice certain parts of the environment and not others, for instance, when we see electromagnetic radiation of wavelengths between about 400 and 680 nanometers (and call it "light"), but not the immense range of other wavelengths, such as "radio" or "microwaves."

The brain that evolved to handle the world in all its teeming complexity was a system capable of integrating simultaneous perception of reality by all the senses. For example, the approach of a bear needs immediate response. This, however, could be triggered by the sight of all or a part of a bear, a sound of running or growling or rustling, or by the smell of the animal. Any or all of these would trigger the kind of hasty departure that was good for the health.

In this ancient, instantaneous-reaction world, events were easily and simply interpreted: a thunderstorm generated the need to find shelter, and fire represented life-threatening danger. For most of the time, most days, however, there was little change in our ancestors' circumstances, so the nervous system developed to make the constant features of the world disappear and only to emphasize the new. This is why, in our natural state, we are geared up for immediate activity or, if not, for a snooze. Gradual change may not be perceived well, but sudden change always is.

Certain elements of perception are fixed at birth: the ability to distinguish wavelengths of light within a certain range (colors); to detect compressions of air that fall between about 20 and 20,000 waves per second (sound); to detect some chemicals with sensors in the nose (smell) and on the tongue (taste); to tell when something is in contact (touch) and when the body moves (proprioception); or to experience certain kinds of physical distress (pain).

We co-opt these senses to help us to navigate the world, to detect danger or communication from others, to avoid physical harm, and to find and choose food. But the senses are flexible navigators. When the world and the signs and dangers out there change, thanks to earthquakes, evolution, or axemakers, the senses also change and look for new things. A hundred years ago we could tell cowdung from horse manure. Today we can tell Rochas from Chanel No. 5. This adaptation to the world as it is begins when we are born because without it individuals couldn't fit into their particular environments.

The way it happens neurophysiologically is surprisingly obvious. Connections in the brain are more widespread at birth than later in life, and the infant's learning does not, as you might guess, increase the number of neural connections in the brain, but instead *prunes* them. The connections that are important to the individual are activated, and the ones that rarely are used eventually atrophy. So how the local environment makes demands on certain connections in the brain (but not others) affects how the brain works and, in a very fundamental sense, how any individual perceives the world. We go with the show.

The first hint of this key environmental role in the development of perception came from cat studies. Kittens raised experimentally for their first few months of life in places where they only see horizontal lines forever after do poorly at seeing vertical lines. Because the kittens saw no vertical lines during the critical period at the start of their experience of the world, their brains eliminated most of the vertical line feature-detector connections. In the natural environment, if a kitten saw no vertical lines in its first few months, it was not likely ever to see any, so the kitten's brain would be modified to detect more nuances in the horizontal lines of its world.

Axemakers, whose gifts change the world, have been running similar experiments on human society for as long as they have been doing unnatural acts like building shelters and cultivating the countryside. The result is that they have made modern Western perception different from others'. In modern Western culture, builders use a lot of straight lines: primarily vertical and horizontal, with straight streets that extend for long distances, rectangular buildings and hallways, square windows, televisions and computer screens. Growing up in this straight up-and-down, left-and-right world has affected our ability to see lines. For example, a study of students in Western cities compared with Cree Indians (whose houses have lines in all directions, not just up and down) showed that city-dwellers are less able to distinguish oblique (diagonal lines) than the Cree. The Cree were similarly not so good at handling the rectilinear.

In another illustration, Zulu people (who live in round huts with round doors and windows and who plough their fields in circles) do not experience a visual illusion called "Muller-Lyer," as much as do Westerners. When we see a vertical line with diverging diagonal lines at the top, we feel we are looking at the inside of a corner, so the vertical line appears to be more distant from us than the diagonals. For that reason, it seems to be longer. If we see a vertical line with diagonal lines bracketing it at top and bottom, we interpret it as likely to be a corner pointing toward us, so the vertical line seems closer and shorter. But if we were like the Zulu and never saw corners like these, we probably would not develop the neural connections that enable us to see the illusion that the Zulu cannot see.

Some of our "modern" afflictions, too, are the result of axemaker gifts. Far more modern people than those in traditional societies have nearsightedness, which is an excessive growth of the eye that puts too much distance between the lens and the retina so that the eye's focal point is above the surface of an object and the image of it is blurred. But since this condition is about 80 percent inheritable, how could it persist over generations, since the 25 percent of people who currently have the condition would not have easily survived to forage or hunt before the invention of eyeglasses? In hunter-gatherer societies, the incidence of nearsightedness is very low, but it is not the case that civilization somehow allowed people with bad eyesight to survive and reproduce. Eskimos were not nearsighted when they first met Europeans, but the first generation of their schoolchildren became so, in the same proportion as in every other society.

The answer to the mystery lies in the way reading at a young age changes the physiology of the developing eye. The "normal" eye gets a great amount of visual stimulation at different distances, but if something in the visual field (like the page of a book) always stays in one plane, the eye grows in only one direction, resulting in focus-changing difficulty. Reading seems to interfere with the eye's growth in this way, since the act of looking at small print in a flat plane reshapes the eye itself and makes eyeglasses necessary. Hence book-worms with glasses.

But it is not only what is "out there" that affects the wiring of the brain. Our physical behavior is also important, and research with monkeys shows that exercising certain areas of the fingertips (to make discriminations in return for rewards) results in an expansion of the amount of brain neurons dedicated to analyzing information coming from that particular area of the skin. This means that when a monkey, or a human, practices a skill or set of movements repeatedly, the brain reorganizes itself to do the job better.

However, although we appear to have *some* innate perceptual abilities, a completely prewired, built-in perceptual system seems unlikely. Human beings have lived in all types of environments in the world, in many cultures, and it is almost certain that much of the perceptual process is developed from experience. Pygmies of the African Congo, who live primarily in dense forest and rarely see across large distances, do not develop as strong a concept of size constancy as we do since they don't see people and animals going off far into the distance. If they are taken out of the forest, they "see" buffalo in the distance as a kind of insect much closer. While this is an extreme example, everybody develops in those ways that will help them to perceive what they, and not others, need.

The prehistoric tools that initially brought about these radical changes in us and our behavior are generally thought of as stone tools, but the majority of the tools of the time were almost certainly made from natural materials that never survived, like bone, horn, sinew, skin, shell, and wood. Two of the most important of these organic tools would have been carrier bags and string. Bags were almost certainly used to carry stones from their original outcrop or to bring meat home from the site of a kill and could have been made from animal skins or from woven leaves. The development of stone tools themselves, especially ones that were used in areas like swamps, where there were no sticks or stones, would have needed some sort of carrier bag. One technology often demands the development of another in this way, as for instance the internal combustion engine would stimulate new road-surfacing techniques, which in turn created rain run-off problems requiring better draining systems. Not to mention air bags and devices to remove pollution from buildings.

Nearly all surviving modern foraging societies have a strong feeling for basketry and cordage. They make nets and traps, play cat's cradle, and have tug-of-war games. String and twine could have been made from animal hide, or from vines or bark, and used for snares, to tie windbreaks together, to make nets for carrying water gourds, or for fishing.

But whatever the tools, perhaps the most powerful and long-lasting change they brought about was that affecting the behavior of the communities using them. The wizardry of making these artifacts conferred power on the axemakers, and in turn on those who could use the tools to do new things. So in a fundamental schism that would last until modern times, the gift of an axe favored those in a community who were good at handling the new tool and the change it could bring. The winners would be those who found it easy to use their minds like that used in making an axe, in the sequential manner. Through the coming millennia, power would very often flow to this analytic type, who could turn the gifts to cut-and-control advantage. It was as if the axe had generated a kind of artifactual environment, in which those who were best at using technology to shape the world (and those around them) became leaders.

This change from "natural" to unnatural selection hastened both the emergence of a sequentially thinking mind and the noncyclic nature of change the axemakers had introduced. These two aspects of human development joined to become a potent force for innovation because the sequential, serial, step-by-step elements of axe construction, when formalized, could be developed into thought processes by which other artifacts could be created. This ability would, as we'll explain below, become a prized asset in the human community.

As a result of this preference, society would elevate science over the arts, reason over emotion, logic over intuition, the technologically advanced community over the "primitive." It may also be that those nonsequential aspects of human talent, expressible, say, in music or art were simply not facilitated in those stringently survival-oriented circumstances and remained dormant, awaiting better days. For the moment, the flavor of the month would be strictly linear thinking.

This selection of minds and the specification of the dominant type happened over a long period of time and did so by the same processes that govern evolution in the natural world: random generation and selective retention. In nature, most things happen randomly. A bamboo shoot turns toward or away from the sun, a frog spawns a new leg, or a new wrinkle appears in the cortex. What happens next depends on the world, which "selects" those changes that suit it. It was Darwin's great genius to see that the world selects in what way life is to be shaped. Lots of sunlight means there will be plants with small leaves turned away from the sun. Less sunlight means those plants that grow larger leaves will predominate.

Inside each of us, as with the kittens mentioned earlier, there are different talents that develop in relation to the world we live in. For instance, people differ in height, but, all things being equal, while a person with genes for more height will indeed always be taller than somebody without these genes, the world they inhabit will also influence how tall they become. Thus for successive generations, Americans were on the whole taller than their parents.

Minds are similarly different. Because humanity evolved from other animals, most recently from the great apes (and before them the old-world monkeys and before that mammals), different talent centers developed at different times in different parts of the human brain. This is why some people seem to be very good at seeing themselves in space and moving in it (talents that are good for moving throughout the wilderness). Others excel at hearing sounds and then matching them to the production of small movements on a musical instrument. Some are good at handling people, or words, or numbers. While individual inheritance is, of course, highly diverse, we are each born with a variety of talents, most of which we never use, because the world does not allow us to. Most readers of this book, for instance, will never know if Swahili poetry, or navigation by the stars, or temple-building is something they'd be good at.

The talents crowd the different centers of the brain, and they include the capacity for sensing the world, knowledge of one's own and other's emotions, the ability to move gracefully, to locate and identify objects in the moving world, calculation, talk, writing, music, organization of self, and others. And there are many more.

The growth and the development of any single person is, like the course of evolution itself, a struggle. Biological evolution is a struggle among different plants and animals, while individual human evolution is a struggle among various talents. Like the kittens who can lose the ability to see vertical lines, we can lose many of our talents as we develop.

In prehistory, when human beings first began to make tools, they changed this "natural selection" process permanently. As in the case of nearsightedness, the axe introduced an artificial change in the way individual talents developed. For the first time, people who were good at sequencing their actions found these talents were in demand, and were rewarded. Those who were good at the process became more powerful, and their offspring stood a better chance of surviving and passing on these talents. But preferentially developing one kind of talent means downgrading or rejecting others. The sequential talents that would bring in meat or construct a village out of the forest would obviously be advantageous, and more and more people would be encouraged to learn these arts. In this way, tools directed the development of minds and vice versa, and over time, this novel, "unnatural" feedback process of ordering and sequencing actions and thought became dominant, thanks to axemaking and what followed it. But we're getting very far ahead of ourselves.

Approximately 120,000 years ago, *Homo sapiens*—sequentially talented, anatomically modern humans—seem to have moved north out of East Africa into the Sahara, living complex lives in rock shelters, building hut encampments when they went hunting, cooking their meat, drying it for storage, and grinding up plants into meal. Some of them seem to have developed cutting tools: a new find at the Semliki valley, in what is now Zaire, uncovered a cache of early spears, cut from the bones of large fishes. Then, over several hundred years the climate cooled sharply and the previously lush, game-filled Saharan plains gradually dried. Some hunters, too far from home to return south in time, were forced to strike out north, following what is now the Nile Valley.

These travelers were surprisingly sophisticated. Archeological finds in Israel (at the Qafzeh cave, close to Nazareth in the Galilee hills, an area that lay in the path of the Saharan emigrants), of objects left behind by these people and dated by thermoluminescence techniques at approximately 90,000 years ago, show them carrying tool-making kits. The kits consisted of saws, planes, adzes, awls, and drills, the kind of things that would have enabled them to produce a wide range of sophisticated and specialist tools. The archeological finds also included tools for woodworking, light cutting and scraping, butchery, boneworking, skinworking, as well as hafts and projectile points.

Clearly the sequential mind was well at work by now, in an entirely new approach to toolmaking. The power of serial thinking to conceptualize something, bit by bit, can be seen in the stoneworking technique called "Levallois" (after the Paris suburb, where nineteenth-century excavations first revealed examples of it). In this technique, the shape of the tool was now determined by the method of preparing the stone instead of being dictated by the stone's natural shape. This ability meant that the travelers could set up their tool-making complexes in a wider range of places. But the real breakthrough in the new thinking can be seen from the way that several tools could now be struck from the same flint. The source stone now produced five times more cutting edge than previous techniques had been able to make. And getting an edge was essential to survival.

The sheer complexity of this kind of toolmaking (and it should be repeated that it was happening 90,000 years ago) has been revealed by modern reconstruction of the Levallois flint-knapping techniques. The most complex tool would have taken as many as 111 blows to shape the flat platform at the base of the tool, followed by one blow delivered with force and extreme precision to split the tool away from the parent stone. Making this kind of a tool indicates remarkable familiarity with the fracturing characteristics of flint. A modern French flint-knapping expert has estimated that a vocabulary of no less than two hundred and fifty signs would have been required to pass on such skill. Also, since each gesture or sound could refer to a tool that was capable of being used in several ways, there would need to be new and different forms of gesture or sound in order to express which use was intended for a tool and by whom.

These "sounds of apprenticeship" may have been the most important ever made by the human mouth. And they may reveal another one of those latent, to-be-selected talents discussed earlier. Gordon Gallup, an anthropologist, has analyzed the limb-movement sequences of tree-dwelling new-world monkeys and noticed "a kind of grammar" in their motions, a succession of actions that must be made in the correct sequence. After the original move out onto the savannah, the underlying brain structure that originally evolved for all that sophisticated, sequential swinging-around was then available for other uses.

So this primeval "grammar" of sequential activity may then have enabled the organization of actions that would make toolmaking possible. And this is where the new power of the serial mind becomes evident. To cut a tool also demands a set of operations carried out in a specific order. The instructions for toolmaking might then have been serial sounds specifying the sequence of physical manipulation necessary to make the tool. The right hand would generally be advantageous in either striking or positioning, while the left hand acted more as a steadying element.

So it might be that the first noises accompanying the "grammar" of sequential toolmaking might also have laid down the basics of the grammar of language, because grammar is based on sounds that only make sense (as do successful tool-making actions) if they are done in the correct sequence. The tool and the sentence would be one and the same thing.

As the tools refined and proliferated, so did the signs and sounds that described them and their manufacture. The member of the group who was master of this vocabulary not only possessed the group's most valuable knowledge but was best able to (literally) articulate it to the community's advantage.

Language would prove to be another, immensely more effective, "axe-gift" with which to cut up and then reshape nature and the community. Initially, it facilitated better organization, more efficient use of group resources, and the manufacture of new knowledge. Ultimately (although the process took tens of millennia), language would prompt humans to become analytical, to segment experiences and reorder them in mental models of reality, which could be used to direct innovation.

As the stock of knowledge multiplied, it produced a multitude of tools to boost chances of survival and get more food out of the environment—needles and awls (in the north, where warm clothing was essential), harpoons and hooks (for the communities along the coastlines), spear throwers and arrowheads (for people hunting the savannah).

Traveling at two hundred miles a year, the human travelers moved out of Africa, and by 90,000 years ago they were in the Middle East.

Fifty thousand years later they had spread across Europe, New Guinea, and Australia. About 25,000 years after that they arrived in Siberia and then crossed the Bering Straits land bridge into America.

With digestive systems capable of taking nutrition from a wide variety of sources, they extracted energy from nature with spear and axe, knife and stone, fire and trap. Each hunter-gatherer needed about fifteen square miles to provide enough food for survival, and this limited the size of their groups to perhaps twenty-five. When they had exhausted the carrying capacity of a living zone, they moved on.

Thanks to their tools, unlike other animals, humans were able to adapt rapidly and survive in widely varying environments. Because of this, seven hundred centuries after they had left Africa as a homogeneous group, these human beings had begun to differentiate. By this time they'd hunted their way across the world and arrived in different climatic zones. Where food resources were plentiful enough for them to remain, they stayed and, after hundreds of generations, had been long enough in different parts of the world physically to adapt to local conditions. So by 40,000 years ago, they had changed into what would become the three main physical groups: Africans, Eurasians (Caucasoids, Northeast Asians, and Amerinds), and South Asian/Oceanians (Southeast Asians, Pacific Islanders, and Australians/New Guineans).

And the longer they stayed in one area, all across the planet, the more they developed different local characteristics, depending on the environment their tools made livable. People who had the tools to survive in dense tropical forest gradually became smaller because lack of sunlight and the scarcity of essential minerals in the soil, due to leaching by the heavy tropical rainfall, limited their calcium intake.

At this time the technology of toolmaking had refined to the point where it was possible to make very small, fine blades that enabled people to sew animal skins, settle around their paved hearths, and survive in cold northern climates near the ice line. Here again the axemakers' gifts were changing our shape, as the northern environment now favored those among the settlers with pale, near-transparent skin, who could maximize the synthesis of vitamin D from low levels of sunlight. Blue eyes that saw better in dim winter light also did well. And colder regions favored heavier, heat-retaining bodies with long trunks and short legs, thick necks, broad feet, and longer, narrow noses to allow the air to pick up moisture and warmth before reaching the delicate mucous membranes of the lungs. Northerners began to look Nordic.

With new tools giving them the means to live where no hominid could have survived before, the travelers were also affected by the level of ultraviolet in the daylight. Successive generations responded by changing the pigmentation of their skin, the form of their physical features, and the shape of their hair. They became tall, squat, fat, pale, brown, yellow, or black and all the shades and shapes in between as the "races" (minor adaptive modifications) began to emerge.

A clear example of this adaptive change occurred among the colonizers who reached Eastern Asia via two different routes. One group, traveling east from Asia Minor, took a route south of the Himalayas. The other group went north of the mountains across Asia. The north-bound group lived for hundreds of generations in the steppes and became physically different from their southern-route relatives, who adapted to conditions in the hotter south to become slim, dark-skinned people able to live in warm, humid conditions, often along the coasts and islands. These people developed bamboo-based technology and eventually colonized Southeast Asia, then settled down as Aborigines and later Pacific Island peoples. The northern group took on characteristics more adapted to their cooler environments and moved on up into Siberia, eventually to become modern Eskimos, some of whom moved farther on, across the Bering Straits land bridge, to become native Americans.

A popular perception of these early humans is that they all lived in harmony with nature in some kind of prehistoric paradise. In some areas that was certainly true for long periods, but right from the beginning human behavior dramatically changed the ecology of large areas, exterminating Eurasian and North American grazing and browsing animals, such as mammoths, woolly rhinos, wild cattle, and giant ground sloths. Slow movers became lunch.

The ice-age people were very efficient hunters of large animals, most likely herding game over cliffs or into lakes, where the animals could easily be speared from wood-framed skin boats. Their use of fire to flush out animals so they could be easily killed also changed the flora of areas of Africa, so the dominant species of trees, shrubs, and grasses became plants that were good at surviving fire, like acacia, lead wood, and wild laurel.

In North America there are intriguing archeological revelations of how extensive the killing was in the form of "ash horizons," where the edge of the area of burned vegetation shows dramatically the considerable distances to which hunting went. Also, by wiping out many large species in an area, the hunters changed the ecology, because those animals were often a key part of the way vegetation propagated.

A typical band of these traveling hunters would number about twenty-five closely related people in the basic unit, and they would keep in regular contact and intermarry with between twenty-five and fifty other similar groups who shared the same language. So a tribe might number anywhere from 300 to 1,000. Before the number rose above 2,000, a tribe would likely split in two and fight. And, depending on the amount of food available, the area a tribe might cover could vary from 200 square kilometers per person in desert areas to 1 square kilometer per person living along a shoreline with abundant resources.

Fifty thousand years ago, as our ancestors were moving into Europe, the environment changed again. Temperatures began to fall, as another great Ice Age set in. These periodic freezes probably occur as an effect of the angle of the Earth's axis (that tilts farthest away from the Sun every 41,000 years) and its varying distance from the Sun (which is farthest away every 100,000 years). When a distant Earth tilts its North Pole farther away from the Sun, the Northern Hemisphere cools down and the glaciers expand. So 50,000 years ago

European weather became catastrophically bad. Throughout most of the continent north of the Mediterranean, the forests disappeared, replaced by stunted brush and bleak, polar desert conditions. It became bitterly cold and worst of all, as the herds depleted and dispersed, food began to disappear.

No doubt as a result of new demands made on them by this deteriorating environment, the axemakers who provided their tribes (and by now, others too) with tools were now working with extreme sophistication, using a new method called the "punched blade" technique. A roughly cylindrical flint core was smoothed and then cut flat across the top. A sharp blow on the edge of this flat "platform" caused a sliver of flint to split off, down the side of the piece. The next blow would then be delivered at a spot near the first. This split off another sliver. The technique could produce up to fifty slivers from one stone core and the slivers (blade-blanks) could later be reworked into specific tool types. Whereas the earlier Levallois technique had yielded forty centimeters of blade edge from one stone, now the same size stone could provide ten meters. The plentiful supply of edges this technique generated made possible a total of no fewer than 130 different tool types. And, as ever, people found a use for them.

The users of this wide choice of edges began, as a result, to live a more complex lifestyle. These northern hunters were now wearing sewn animal-skin parkas, they lived in open country in summer and in river valleys in winter, and they carried their portable fires with them as they moved. Every spring they would come out of their winter caves to return to the same south-facing hunt sites, where they put up rectangular tents of hide with paved pebble floors and, in some cases, circular hearths. They hunted with spear throwers, with the spears themselves held on fiber lines, and with detachable blade spearheads so that the valuable haft could be retrieved. Barter with other groups had begun, too, bringing in artifacts from as far away as 250 miles.

That they also buried their dead adorned with shell necklaces, pierced mammoth-ivory beads, bracelets, head bands, rings, and finely crafted flint blades means they were dressing them for the afterlife. Many of the dead buried in this way were children, who could not yet have earned the kind of reputations that would earn them this special kind of burial. So it may be that these were members of families, or children of particular men or women, who held power. These dead children may indicate the presence of an elite, whose position was possibly hereditary and who held enough authority or possessions to order extremely valuable and magic articles to be placed in their children's graves.

Around 30,000 years ago, as the temperature continued to fall, imperiling food supply, and survival began to require ever-more efficient kinds of organization, there was an extraordinary change in the behavior of the inhabitants of a swathe of southern Europe stretching from Spain to Southern Russia. They began to create the first art.

This art may be the first, indirect evidence of the myth-making use of their new language. It was possibly used by tribal shamans as a tool for social control, in the form of magic explanations for natural processes which only the shamans knew. The authoritative nature of these explanations might have conferred magic power on the shamans, who used their mysterious knowledge to predict natural phenomena. The art may have been used to provide ritualistic settings for ceremonial occasions. It occurs first in caves, probably sacred places, on whose walls the shamans and their assistants painted images of animals, and where they held initiation ceremonies (in some caves the hardened mud floors show evidence of dancing feet). The purpose of the paintings seems to have been to placate the forces of nature on which the community depended.

The bison, horses, lions, and reindeer in the paintings were all targets of the hunt. Some of the paintings also feature hunters attacking wounded animals with what look like numbers of spears stuck into them. These may have been drawn during rituals, as symbols of the successful hunt to come, at the start of the hunting season each year, and intended to give the hunters strength of purpose. But whatever their purpose, the pictures were obviously hidden from the eyes of ordinary people because in some cases they are found deep underground, in caves within caves within caves, suggesting that the journey to reach them held exclusive, ritual meaning.

Cave art comes at a time when a rapidly growing Upper Paleolithic population was living in tough times, which demanded constant adaptation and resourcefulness. The large and growing number of new tools produced in response to this need may have complicated the community structure, perhaps as new tools made possible new and more specialist activity. The consequent need on the part of tribal leaders to hold such an increasingly heterogeneous group together under increasingly difficult conditions may have in turn given rise to the need to find a more powerful source of authority, even greater than the leaders.

The paintings found at the *Trois Frères* cave-sanctuary in southwestern France include the representation of a half-man, half-deer, named by archaeologists "the Sorcerer," and this may be one of the earliest figurations of the new authority: a god who held the well-being of the community in its power and with whom intercession could only be performed for the leader by the shaman. In the constantly changing environment of this time, the introduction of this kind of supernatural mythology could have made the command hierarchy more effective and consolidated the unity of the group in the face of the strains placed on it by the fear that the weather could get even worse and survival more precarious.

As the glacial weather caused the game herds to disperse and it became essential to be able to monitor their movement over wider and wider areas, it would have also been natural to want to rely on other groups for help or information, and group alliances through marriage links would cement those relationships. This may have been the reason for another new artifact that first appears about 20,000 years ago. It's a small, carved, female figurine known as the "Venus." The figure appears increasingly throughout Southern Europe, in an area stretching over a thousand miles from western France to the central Russian plain.

Venus has a uniform shape and probably acted as a kind of identification, carried either by people who were likely to make contact with other groups, or more likely by women given in marriage alliance, to remind the adoptive community of their origins and in this way to ensure continuity of contact between the groups. As separation brought increasing language problems, these identity-artifacts might have helped to avoid miscommunication if long-distance hunters or traders meeting others found it difficult to explain who they were. The artifacts would also have made possible the maintenance of intratribal links at great distances, enabling groups to scatter over very large areas.

By this time, the brains of these trading, traveling, marrying tribes were anatomically fully modern. Endocasts made inside the remains of their fossil skulls indicate a major increase in the supply of blood to the brain, as well as growth in the size of the Sylvian fissure, which is related to language production. Broca's area, only present in the highly complex brain of modern man and associated with speech, also first appears in these new skulls.

It was around this time that an extraordinary new kind of artifact first appears. It is also another powerful example of the way in which the axemakers reshape the way we think. The new tool must have seemed entirely magical, and it is tempting to see in it the origin of the age-old myth of the magic wand. It seems to represent the first deliberate and detailed use of a device which would serve to extend the memory, because with it knowledge could be held in recorded form outside the brain or the sequence of a ritual. These magic objects are referred to by modern archeologists as "batons," and they are made of carved bone or antler horn. Several thousand examples have survived, and they appear in most of the cultures of the period.

Each carved mark on the baton is made with a particular type of tool stroke. Some are simple line marks, others are curved, others look like dots. The marks are also made in sets, each set placed horizontally in a line. In some cases, the engraver has turned the bone over and continued the series of marks on the back, in order to find room for all the marks. This alone would indicate that the work is not intended simply as art. In all likelihood the carvings represent the first form of information notation. Their existence alone is evidence of the highly developed stage of their maker's intelligence. The cognitive faculties needed to make the batons required a brain capable of a complex series of visual and temporal concepts, demanding both recall and recognition. These are exactly the same mental abilities which are involved in modern reading and writing. So the artifacts reveal the presence, approximately 20,000 years ago, of fully evolved, fully modern, brains. But the way these brains thought about the world was still very different from ours.

The batons conjure up an existence filled with magic symbols like the Venus figurines, with rituals associated with cave art and life beyond the grave. The baton notations fit within the context of a sophisticated cultural repertoire that included decorated tools, red-ocher-painted amulets, personal decoration, ritual objects and images, and grave-and-burial rituals involving carefully arranged funerary objects with flower arrangements and anthropomorphic representations, such as the stag-man "Sorcerer" mentioned before. The members of these communities were very far from the shambling, apelike cavemen envisioned by modern archeologists until only a few decades ago.

The first clue to the batons' specific purpose comes from the fact that they are all found in the southern latitudes, around the Mediterranean, principally in France, Italy, and Spain. It is here that, as the ice began finally to retreat, climatic improvement would have been accompanied by more vegetation, an increasing variety of game, and the opportunity to benefit from both. The second clue is that the regularity and repetition of the carved pattern sets on the batons indicate a sense of periodicity.

As time passes, the number of carved bones increases dramatically, and the sets of marks begin to include representations of animals and plants. The Montgaudier baton (a carved reindeer antler dating from 17,000 years ago) carries lifelike engravings of seals and fish, even though Montgaudier, in France, is a hundred miles from the sea. Microscopic examination reveals why. The carving of the salmon jaw shows the shape characteristic of the fish at spawning time. The same baton also carries images of serpentine forms that might represent local snakes coming out of winter hibernation at the same time as the salmon spawn. A budding spring flower also on the baton completes the carvings and reveals why the engraver drew seals and salmon so far from the sea. In spring, the salmon would begin their spawning run up the rivers and predator seals would follow them. Both these animals were a plentiful source of food, and the batons made possible an accurate prediction of the time of their arrival.

A later, carved bone, found in Cuerto de la Mina in northwest Spain, develops the theme, its two faces carved with a series of images showing animals and plants in a seasonal sequence running from March to October.

The most extraordinary example of all, the French "La Marche" bone, dates from 13,000 years ago and was discovered in a find that included decorated tools, amulets, and a library of stones engraved with human and animal images. Apart from a carving of a pregnant mare, the bone carries a series of marks carved in sets and subsets, each set made with a different tool.

When compared with an astronomical model, the bone is revealed to be an exact sixty-mark notation of the lunar calendar. The subsets begin at conventional lunar phase points, within the observational limitations to be expected in the European mid-latitudes. The entire calendar covers a period of 7.5 months with remarkable accuracy. The whole sequence seems to run from the beginning of the thaw in March to the first frost in November, covering the period when the hunters would be able to live out of their cave shelters.

These wonderful batons indicate the ability to abstract and symbolize. They also reveal a highly developed capability to observe and record celestial phenomena. Above all, they illustrate how the tools, which were making possible an increasingly complex life, also changed the way in which our minds worked. As well, the use of the calendars to organize a hunt indicates the capacity to plan strategically over an entire year and to express these plans intelligibly to the other members of the group. Above all, the group's ability to comprehend all this information indicates that the symbolism involved in explaining it was shared. This level of communicative ability would in turn make possible much more rapid cultural adaptation.

The key significance of the batons for the future of the human community (and why they show the power of tools to shape the mind) lies in the way this external memory device increased the working capability of the brain. A tool like the baton permitted the codification of nature into durable symbols that could be used and reused by the imagination to manipulate the world. With it, the mind could symbolically dissect the world, chop it up, and then rearrange the bits, to find new patterns in the data. In this way, symbols gave the user the ability to run scenarios, to see outcomes in theory before committing to practice. So the baton gave the shaman —then the repository of arcane knowledge (and through him the group's leader)—the ability to predict events before they happened, like the thaw, or the arrival of the salmon. The success of the new tools is evident from the fact that, with very few exceptions, the batons all show signs of continuous use.

It may not be too fanciful to imagine how, as the travelers continued their millennial walk around the planet, they survived and multiplied, thanks to these portable data banks with their seasonal and hunting data interpreted at magic moments by the shaman.

But above all, the very presence of these magic wands announces a new kind of knowledge that was perhaps unlike what had preceded it. The baton was not just a flint axe shaped by mysterious skill, unknown to the majority, but whose purpose could be seen in its use. The symbols on the baton were visible, but they were, to all but a few, incomprehensible under any circumstances. No amount of looking at them or touching them would have made their meaning clear without the special code that only the shaman or his acolytes knew. The symbols were visible proof of the existence of a kind of artificial knowledge of the world which gave power to those who knew how to use the knowledge. It was the kind of knowledge that would increasingly come from the axemakers and that would immensely widen the gulf between those who made change and those who merely accepted it.

The batons perhaps also caused one last effect, cutting deep into the minds of these members of the early human race. The language they had spoken as they set out from Africa was shared and must have been sufficiently developed to describe the proliferating tools and their various uses, as well as to organize the social complexity they caused. But as the success of the tools themselves helped the travelers to move farther and farther afield, the groups would separate again and again as they took separate paths, down different valleys, along different rivers, into different mountains, so as to survive more readily apart than together.

And as time passed, and the human race split again and again, those original moments of farewell, somewhere in the Near East, possibly became a half-forgotten memory recalled only in myth or ritual. So too did the common language we had all once shared. As tools became more environment-specific and as the noises used to describe them (and everything they made possible) became more exclusively linked to locality, we lost our once-single identity in a babble of new dialects, which over time became different languages operating in differently organized brains. The axemakers' gifts had given us different ways to express different realities and different views of the world based on environmentally generated different value systems.

By 12,000 years ago, the now physically and culturally diverse tribes were scattered in mutual incomprehension through every continent but Antarctica, their common African ancestry forgotten, their existence firmly rooted in the lands to which their tools had taken them. They could not go back anymore, now. They could only stop and settle.

Chapter 2

TOKEN CONTRIBUTION

Man's development and the growth of civilizations have depended, in the main, on progress in a few activities—the discovery of fire, domestication of animals, the division of labor; but, above all, in the evolution of means to receive, to communicate, and to record his knowledge, and especially in the development of phonetic writing.

COLIN CHERRY, *ON HUMAN COMMUNICATION*

About 12,000 years ago, when there were about five million human beings on Earth, the axemakers produced two gifts that would bring extensive change to the physical world in which we lived and to the landscape of our minds. Our ancestors took the gifts not long after they had ended the great journey, because, as was to happen so often throughout history, they had no choice.

The tribes had survived during the millennial trek across the planet because of their tools. And thanks to the way those same tools allowed them to derive more and more subsistence from nature, their numbers had grown to the point where there were now too many of them to continue to survive without another radical change in behavior. The new gifts of agriculture and writing would free them from the vagaries of natural food sources and take their shaman bone notation to a new and world-altering stage.

The travelers had for some time been using tools to enhance the carrying capacity of the places in which they found themselves as their numbers made this increasingly necessary. Part of this process was very likely the development of a kind of protohorticulture. Foragers have to be keenly aware of the seasonal movements of animals and the life cycles of plants, so they could well have noticed that regularly eaten plants would be more likely to grow near temporary settlements where the seeds had been discarded. They might well have protected the plants they particularly enjoyed from animals with traps or snares, or even encouraged growth by doing weeding. In hard times, the more productive areas would be carefully guarded from rival tribes.

So when it became clear that settlement was likely to be the only way to survive, it must have been in conjunction with the ability to make sure survival was possible without having to travel too far for food. Evidence shows that tools to make this possible were already at hand. Primitive sickles and grinding stones in Israel as early as 15,000 years ago are evidence that the hunting groups were by this time tending toward a more sedentary life. The new tools enabled a more efficient cropping of food from the local environment, so the population increase continued. And as their tools began to tie them more specifically to their now semipermanent location, the groups also began, according to their circumstances, to rely on more location-specific activities: shellfish collecting, grassland hunting, or forest harvesting. Behavior and location began to be intimately linked.

About this time, burial practices began to be adopted for more than just the leaders, which may indicate that personal names, previously only used for chiefs and shamans, had begun to attach to every member of the group. Names, perhaps chosen by the leaders, would bind the individual to the group both before and after death, and this in turn might strengthen group identity and, by implication, the power of the leaders. This would also perhaps have ensured that people were likely to remain with the new settlement.

Perhaps the first tribe to settle permanently were the Kebarans, who did so on the plains of the Levant, in view of the Mediterranean, at a place that was fertile with plentiful fish and fruit. The culture that followed them (known as "Natufian" and located in what is now Syria) developed, and as it did so the size of settlements increased dramatically from at most four families living together 11,000 years ago to the first villages, consisting of more than two hundred houses (in Mureybit, in Syria) two thousand years later.

But the new Natufian settlers did not become instant farmers. They were still expert hunters because they had another new tool, a long piece of basalt in which were ground two deep, parallel grooves. When the stone was heated, wooden shafts could be easily straightened in the grooves. This would greatly improve the accuracy of arrows and make it more likely that hunting could provide sufficient food for a settled community, even if plant sources failed.

Initially, as population growth forced the wandering groups to seek more reliable sources of nourishment, the limited carrying capacity of the land also increased the value of the ancient plant lore of women gatherers and their knowledge of dry agriculture. The introduction of a dry-farming scratch plough (basically a digging stick pulled by an ox) also radically improved the output of the local environment, although it was at first inadequate to support a healthy, well-fed population. Diet was restricted and malnutrition initially common. But where it had once taken fifteen square miles to support a hunter-gatherer, a settler now needed only three.

The basic diet of the modern world comes from this time, when our early ancestors chose local species for cultivation: emmer wheat, einkorn wheat, two-rowed barley, peas, lentils, broad beans, and vetches. These are found from Iraq to Kurdistan, Palestine to Western Turkey, and from the Anatolian Plain to the Levant.

Also, as the new agricultural techniques spread and diffused from the early centers of innovation, they encouraged linguistic consolidation among the settled groups using them, and this strengthened the sense of identity with their own techniques and traditions. This cultural stability gave rise to the final, settled form of the major modern language families: Indo-European, Afro-Asiatic, Elamo-Dravidian (India), Sino-Tibetan, and Austronesian.

And in one of the last physical changes to be caused by tools, agriculture removed the need for big teeth. When cereals replaced meat, teeth became smaller and the face became more vertical. Evidence for the dietary switch can be seen in the number of Near East rectangular granaries surrounded by soil containing dense amounts of fossilized cereal pollen.

As the settlers' numbers continued to rise, they moved to more fertile locations by rivers and on coastal plains. In their new settlements, they lived in tiny villages of mud brick houses with reed roofs, separated by narrow alleyways. Their tools were more diverse too, as flint and stone bladelets gave edge to arrow points, sickles, and borers. They sewed skins with bone needles, processed their cereals with stone pestles and milling stones, wove reed baskets and mats, and herded sheep and goats.

The food surplus generated by dry-farming had by this time established an economy that could support people who did not contribute directly to the production of food: craftsmen, scribes, doctors, chiefs. Each one of these possessed some form of esoteric knowledge that the general population did not share, but which it needed. Decision making and social responsibility became more and more concentrated in the hands of these new specialists.

So by the time the wandering tribes had finally settled, they had the tools with which to feed and shelter themselves far in excess of anything their ancestors could have enjoyed. Then, some time round 7,000 years ago, either the population increase or a harvest failure, triggered by the increasingly long spells of dry weather, drove some communities, probably near one of the great rivers, to switch from dry farming to irrigation. It was by then part of traditional lore that plants shed their seeds and that in well-watered areas these seeds would grow. The breakthrough in thought was to understand that these natural processes could be reproduced artificially.

Food surplus must have come soon after deliberate irrigation, or even the simple planting of seeds in naturally watered areas. The event seems to mark a watershed in the attitude toward the environment. For thousands of years, humankind had been intimately involved in nature. For hunter-gatherers, the natural world was a living entity, with the seasons providing different kinds of food and shelter. And we must have been highly aware of its every nuance: the location of a fruitful autumn clearing, the place where birds could be netted in spring, the springs with the clearest water, the rock shelters where the winter winds did not penetrate. Above all, perhaps, we knew the essential relationship between food and timeliness, whether with seasonal berries that appeared briefly and then were gone, or with migrating animal herds that did the same. If you missed nature's cues, there was no second chance that year.

The first agricultural surplus changed all this at a stroke. Nature was now reproducible, to be cut up and controlled at will. And with this new concept came a parallel requirement to apply the same cut-and-control methods to the community itself. There might well have been enough surplus food to support a more complex society in the growing villages, but since the population was now too numerous merely to be able to walk to new sources of food, as their ancestors could always have done, survival from now on would depend on being organized in ways that had never been needed before. For this reason we would begin to feel the need to stay where we were.

For the first time, thanks to the axemakers, we were about to live in "places" from which some of us would never again move. We would think of ourselves as "of" those places. In the form of large villages, they would become our "home." We would, from now on, identify with one location and the people with whom we shared that site. Together we would be "of that place" and others, similarly settled, would be from "their places." Our village walls would mark the space within which we were what we were, doing things differently from others living within other walls.

In these extraordinary, new, artificial enclaves, we would no longer be a passive part of nature. Even the way we conceived of direction must have changed as the natural features indicating north or south, east or west became something permanent in life, rather than a changing reference to seasonal winds or to the movement of the sun and stars. In a sense, the very world of the human community had been defined by the tools that had made the new settlements possible.

That new definition was now made visibly clear with the second of the two new axemakers' gifts. It would make possible higher levels of organization required to keep the community viable and to help it survive. But organized survival would also require new levels of obedience, new constraints on behavior, new layers of social authority. The new gift would eventually make us think in a new way. It was the gift of writing.

The act of cutting into a stone to make tools now became a

tool for reproducing the world in symbol. The first writing was a new, improved version of the shaman's carved baton (by now no longer capable of handling the complexity of the information base required by a more populous community), and it provided early agricultural communities with a new way to describe and record the world.

The new technique would provide a radically different means of generating knowledge, an unparalleled way of manipulating information external to the mind, and, most effectively of all, a medium by which social control could be ever more rapidly and powerfully enhanced. Although it began around 10,000 years ago, the full development of writing would eventually take 7,500 years and then would remain essentially unchanged until the cognitive revolution occurred in first-millennium-B.C.E. Greece. But first we go back a bit, to look for its origins.

In the beginning was the bottom line: numbers, not words. First of all, the availability of surplus meant that there was more food than the community required. This surplus might be saved for later use. It might be used to pay for services by those people not primarily employed in the matter of food production. It might even be used as gifts or as contributions to religious ritual. Whatever the case, its existence required inventorying, and that required measurement.

The intellectual leap expressed by the development of counting that took place during this period is like the leap in the cognitive ability of a child during its early development. Early on in their life, children reckon in "ones," "twos," and "many." This ability is probably the basic human understanding of quantity, since it is found in children of all primitive and ancient societies. A Vedda tribesman of Sri Lanka who is counting coconuts will assign one stick to each fruit, keeping a correlated record: one coconut, one stick. Ask him how many coconuts he's counted and he will likely point to the sticks and say something like: "that many."

The same is true of all early societies we know, where, for instance, to account for a large number of animals (say, fifty-eight sheep), each one would correspond to a marker, like a pebble. A marker for each sheep would be set down at the beginning of the day and picked up again when the herd was rounded up in the evening, when they and their shepherd went home. If no pebbles remained after the sheep had been herded, the shepherd would know his herd was complete without having ever to know how many sheep he had.

Representing and reducing the world to abstract symbols and numbers forms a major element in our modern way of thought, but it is not part of the natural complement of human talents. Speaking and listening come naturally to almost all human beings, but writing and reading are difficult to learn. The prehistoric innovation of representing quantity and then words was something that developed over a culturally-long, though biologically-short, period of time. In terms of counting, for instance, the set of social changes that prompted human beings to represent the world of objects and quantity by abstract markings (and in this way to distinguish three objects from four and later "three" of *anything* from "four" of *anything*) probably took about 10,000 years.

The first genuine writing arose approximately 12,000 years ago, in the Zagros mountains of Iran and in Turkey, because harvest and herds, now surplus to immediate requirement, were property and demanded some kind of a mark of quantity and ownership.

The first examples of writing occurred in the Near East at the time when animals and plants were being domesticated. They took the form of clay tokens, smaller than an inch in size, used to represent different commodities. A cylinder stood for an animal, cones and spheres represented, approximately, pecks and bushels of grain. Each token stood for a specific quantity, although the shape of each token represented a different item. Two measures of grain would require two tokens, fourteen measures would require fourteen. Between 10,000 and 5,400 years ago, token use gradually spread all over the Middle East. Token shapes were standardized early, possibly indicating mass hand-manufacture, and they may also have been the earliest artifacts to be kiln-fired.

These tiny ceramic objects were the genesis of written language since each token was a separate unit of meaning. Each was discrete, their use was systematic and abstract, and used together they followed an order or syntax. This habit of systematic usage certainly laid the cognitive groundwork for those further developments in language- and mathematics-processing that would later develop into an organized, written, language and number system. These first tokens were like the letters of a new kind of alphabet.

Their most immediate contribution was in counting and accounting. Each token had its specific sign and also stood for a fixed quantity so, as stores grew in volume, more tokens were needed to represent the greater quantities involved. To keep their tokens together in individual transactions, the Sumerians of Mesopotamia (in what is now Iraq) placed them inside clay "envelopes."

All the tokens that have been discovered also carry the personal seal of an official. Finds of such tokens in graves only of high officials indicate that perhaps, in association with ownership of property, the new artifacts came to be regarded as symbols of status, perhaps even of hereditary office: they were tokens of power.

These relatively simple, though numerous, tokens counted and accounted a world of stockyards and storehouses. But the complex urban world now forming needed even more complex representation. Tokens now began to carry additional markings such as notches and patterns that indicated new commodities, like perfume, bread, woven articles, and clothing. Finally, reflecting the variety of goods available, there were fifteen token types subdivided into no less than 215 subtypes.

The increasing number of token types soon made the practice of using the clay envelopes too cumbersome. Since envelopes made of clay are not transparent, in order to validate the contents an envelope containing tokens had to be broken. This minor inconvenience triggered a major event in the history of information storage that was to create another new kind of "knowledge," which would, as ever, be restricted in its availability and use.

In all probability, the event took place somewhere in what is now Syria or Iraq. To make things easier, somebody had come up with the idea of pressing tokens against the wet clay of the outside of the envelope to indicate the number and type of the tokens inside. At some point, however, five or so thousand years ago, somebody else seems to have realized that it was even easier to do away with the tokens inside the envelope and use only the impression on the outside. In time, the envelope itself, now empty of tokens, was flattened into the form of a tablet carrying the token symbols.

Given the growing number of people and commodities, the new technique must have stimulated more attempts at representing information, now that marks were being accepted in place of solid objects. At about the same time, in another significant advance, the first arithmetical signs appeared in the form of signs for quantity. Where previously three sheep had been represented by three individual discs carrying an incised cross (the one-to-one correspondence described earlier), the Sumerians now produced a separate abstraction to represent quantity as a number.

They first used the number symbols to represent measures of grain, and because grain was the staple commodity, the symbols for number were understood by everybody. Then, about 5,000 years ago, a breakthrough occurred. The accountants of Uruk, one of the first towns in Mesopotamia, were able to abstract, from the concept of "two olives" and "two sheep" and "two sheaves," the conception of "two-ness" independent of the individual objects in question. These accountants devised twin signs: numerals that were specific numbers, and pictographs that expressed commodities. They were produced differently, numerals being impressed into wet clay and pictographs cut into hard stone. One tablet from Uruk shows one of the first depictions of this with five sheep described by the pictograph for sheep, coupled with five impressed wedges.

Then they improved the system. A line ("small") now meant "1" and a circle ("large") now stood for "10." And these could be combined. One circle and two lines stood for "12." At first, the signs were applied to grain payments, then to the number of workers to be paid, and finally, as proper numbers, they were used to stand for quantities of any kind. Used as a means to manage and account for inventory and the movement of goods and animals, these token marks are evidence of further advance in control over nature and over the community. Number and commodity were now cut apart forever and, as a result, most significantly, numbers could now be applied to quantify any thing around in the world. We were now able to think of the world as something which could, like grain or sheep, be inventoried, controlled, and redistributed.

Around this time, as the population centers increased in size and number, new axemaker gifts facilitated their organization and maintenance. The ox-drawn plough boosted grain production, the wheel and the sail transported it, the potter's wheel made jars to store it, and the waterwheel ground it into meal for people now living in houses made of kiln-fired bricks in communities protected by metal weapons. Draft animals fertilized the soil, the plough increased the area of workable land, and "short-fallow" farming (with frequent cropping and growing) produced crops in faster sequence. Things were changing faster now.

The widespread establishment of agriculture marks the point when axemakers' gifts had given us the ability to change our environment within the space of a single season and reduce the length of time a community would need to survive after a bad harvest. Only a few thousand years after the invention of irrigation in the Middle East, the Mesopotamians were making the desert bloom and changing the character of the land in ways nobody had ever done before.

This "hydraulic" civilization, emerging around seven thousand years ago, brought for the first time the realization that humans could make large-scale changes in the shape of the natural world, as networked irrigation canals distributed water from the rivers to nearby farmlands. This new ability was celebrated in the dominant theme of all Mesopotamian mythology: that the chaos of nature could be transmuted into a human-divine order. Society and its environment were now similarly controllable.

The change to a sedentary, agricultural society had also radically changed the role of women. Earlier, it is likely that women's skills included knowledge of the ecology of food gathering. They probably guarded the fire and knew how to make wooden and clay containers, as well as to cook and use animal parts for utilitarian purposes. They likely knew herbal medicine and could turn plants into dyes and yarns, and weave them into clothing. So their skills were equal, if not superior, to those of men. But when agriculture provided surplus property, this bestowed on its owners, who were almost exclusively men, the power to distribute the surplus. The association of ownership with the acquisition of territory by use of arms almost immediately began to exclude women from these positions of gift-giving power.

The growing community surplus was by now big enough to support a wide variety of crafts. The town now included herdsmen, ploughmen, oxmen, fishermen, butchers, brewers, bakers, boatmen, farmers, gardeners, builders, carpenters, potters, and weavers, as well as those dedicated to the production of luxury items, like jewelry and oil lamps.

In spite of this, each town or village was also, perhaps, unable to survive on its own resources. For instance, although Mesopotamia (approximately modern Iraq) was abundant in cereals and livestock, the flat alluvial land, with few major outcrops, was deficient in minerals. So it became necessary to build up a reserve of food or craftsmen's output to use as barter for surplus food or materials from other communities.

And now, driven by population growth, generated in the first place by axemaker tools, competition for resources demanded a new kind of leader: one who could command in war and in peace and organize redistribution of food and materials among his people. To do these things, aided by the emerging, though tiny, elite of pictograph readers and writers (perhaps a few dozen in a community of thousands), the chiefs mobilized labor and used tithes to support even more specialists, including the smiths and metalworkers who could produce and maintain military materiel.

For this reason, perhaps, the new leaders, whether religious or secular, are regularly referred to as beneficent givers of goods, generous and bountiful. In fact, the process was much more the other way. The leader collected tithes and taxes, so as to have material to distribute in return for the labor that went to produce the materials themselves. His prime, self-ordained function was as organizer and protector of the rapidly growing communities now too large to be called villages.

The largest, and perhaps the earliest, Mesopotamian city was Uruk, which came into existence during the fifth millennium B.C.E. (seven thousand years ago) as two separate settlements, each on either bank of the Euphrates. During the next millennium, the city retained its twin-site form, but now it had two prominent ceremonial districts, one dedicated to the sky god Anu and the other to the goddess of love, Enanna, while its area increased from about 25 to more than 175 acres.

Uruk grew dramatically during the century or two following 3100 B.C.E. At first the size of the city rose from a hundred to two hundred and fifty acres while the number of settlements outside the city walls increased from about a hundred to a hundred and fifty. Then came very sudden change. The city continued to grow at a faster rate than ever, but it was now growing at the expense of the countryside. About half the settlements in areas close to the city were abandoned in a process that caused Uruk to grow to about 1,000 acres and its population to double from 10,000 to 20,000. This was about two-thirds of all the people living in the region. The speed with which this growth occurred suggests that the population felt an urgent need to concentrate, and the most likely reason for doing so was an external threat.

As the community grew, people were organized into what to the modern view seems an oppressive and claustrophobic regime, a new social network based on a new, society-wide hierarchy, resting on the now varied and sophisticated axemaker gifts of agriculture. At the basic subsistence level there were farmers and their families, working long hours to produce more food than they needed. These people carried the surplus to intermediate distribution points, where officials made sure the right proportion of the food went to the largest centers of population, and an extra-large share of the best food went to people in authority. In return, the workers were very probably paid with mass-produced pottery and textiles and, above all, assured protection. Religious rituals in support of the rulers would also take up much of their time.

But the way these conditions were accepted indicates the new view of the world brought by the axemaker's gifts. The rewards of living in cities far outweighed the behavioral constraints urban life imposed. To those living outside them in country villages, the cities must have seemed centers of almost magic activity, with their walls and tall buildings, peopled by small numbers of men who could read and write, patrolled by others carrying weapons of precious bronze, ruled by mysterious figures who were supposed to be half-divine. The cities were, in every sense, the first super-powers, and a great deal of pressure must have been needed to persuade country folk to remain outside them, down on the farm.

The position of "king" that emerged around this period in Mesopotamia may well have originated late in hunter-gatherer times, when the ritual formalization of myths generated religious practices that featured a semisacred shaman. Early on, the governing elite in Mesopotamia consolidated its position by associating itself with the mysterious source of power the shamans had invented. With the shamans' help, only the kings now understood and could predict what the forces of nature might do. The new leaders presented themselves as mediators between the population and the ancient mythical forces, and claimed direct, divine contact with these supernatural forces, anthropomorphized early on as gods and goddesses.

So the royal leaders were the only ones who could intercede with the deities and ensure their continued beneficence. The divine nature of these early heaven-talking hierarchs is indicated in Sumerian pictographs, where the scribe has placed a star symbol before royal names, indicating their link with the cosmic beings in the sky. This sacred and privileged position is mirrored in a new tendency for leaders to amass wealth and to take it to heaven with them when they died.

The concept of a new kind of human being, in the form of the high, controlling authority, separate from and elevated above the populace (a concept still very much alive in the modern world), was matched (as it is today) by the amount spent on buildings to house these people and their staff. The new power elites were not, as semidivine beings, expected to live with the rest of the community, so ceremonial structures and the houses of rulers became larger and more prominent, sited on hilltops and closed round by massive walls. The rulers' skeletal remains show that their privileges ensured them better health and greater longevity.

From now on, the trappings of leadership were to be public symbols reflecting the permanence of the community and the new values imposed on it by authority. The special status of the royal family is reflected at Uruk, where the temple and its mound, standing forty feet high and covering an area of 420,000 square feet, its buildings and walls of plastered mud-brick covered with patterns of thousands of clay cones, is estimated to have taken 7,500 man-years of effort to build.

The stringent social control required to make the Mesopotamian cities function made rituals and artifacts extremely conformist. At Yahya, not far from the city of Sumer, was a center of stone bowl production. These bowls were luxury objects, produced for the Sumer authorities (since they were not used by the locals at Yahya, and there is no evidence of a local elite) in return for food and commodities. Bowls carrying the same symbols as those at Yahya have been found as far away as Syria and the Indus valley. These luxury articles acted as images of authority, perhaps much in the same way as modern artistic treasures in city museums and national institutions do today, because only the authorities could afford to commission them.

By this time, the abundance of personal seal impressions on the flattened clay tablets indicates a complex and well-established trading system. Goods were passing through enough hands for forgery to be a matter of concern, and seals made forgery less easy, as the seal was typically worn on its owner's wrist. But as number and commodity pictograms became too cumbersome to keep up with the pace of life, pointed reeds, previously used to draw images on wet clay tablets, were discarded in favor of a stylus. This could be used to mark a line in clay with one movement, and since the cross-section of the stylus was wedge-shaped (cuneiform), it gave its name to the now stylized, fast-written pictograms.

The new gift of writing by this patterned process was still, however, extremely complex and esoteric, consisting as it did of at least 2,000 different signs. There were, for instance, no fewer than thirty-one versions of the sign for "sheep" in various states, conditions, and types. Mastery of these signs required many years of training and made the art of writing a highly specialized skill known only to a few. The inaccuracy of the phonetic renderings of signs (representing in many cases closely similar sounds) as well as the confusion created (because the same sign often could be read either as the picture of a sound or of an object) all made the learning process lengthy and difficult. While writing may well have facilitated the development of a much more complex and heterogeneous community, giving many of its members a style of life undreamed of only a few generations earlier, the rewards came at a cost. The power of writing to organize and command was still inaccessible to all but the very few.

The scribal bureaucracies maintained control of taxation and the allocation of resources, of workers' remuneration, and of domestic and foreign trade. Later, in Egypt, the representations of economic activities on monuments would invariably show a scribe in a prominent location, keeping track of the operation. Because of their importance in the administration, the scribes held a respected and privileged position, so they had no interest in simplifying their arcane art, since that would serve to make it more accessible to potential competitors.

Administration of such a system required a large, specialized, professional clerical establishment. The scribal school, or *edubba,* was the driving force behind society and generated a small, literate elite. The schools came into existence early in the third millennium, and the full-time course lasted from childhood to early adulthood. Study began with practice of the syllabic characters, such as *tu ta ti, nu na ni, bu ba bi,* etc., after which a repertoire of some 900 signs was taught, and this was followed by groups of more than one character. Following memorization and practice of these fundamentals came the real reason why it was difficult for many people to share the knowledge of reading and writing. The student was confronted with months of study in order to learn lists of thousands of items arranged by subject matter, like parts of animals and the human body, names of domestic animals, of birds and fish, plants, implements, etc. The difficulty with pictograms was that there were almost as many sign patterns as there were things to be signed.

In their literary, linguistic, mathematical, and astronomical achievements, the scribal schools far exceeded practical and bureaucratic needs, and in doing so (small wonder) widened the division between the leaders and the led. The schools, new arenas of intellectual activity created by axemaker technology and closed to the majority of the population, used literacy to develop an unprecedentedly comprehensive, though extremely selective, educational system, and through it they established a powerful elite. Many of the thousands of surviving tablets give the names of the scribes and even the names and occupations of their fathers. As might be expected in a society where literacy was a jealously guarded skill, scribes came from important and wealthy families and many went on to high office in the administrative bureaucracies.

Then, in a process that would be repeated again and again throughout history, the qualified few were obliged by changing circumstances to diffuse some of their specialized skills. As technology began to generate increasing social change, the unavoidable alternative to social collapse was to allow a wider segment of the community to read and write. The number of pictographic signs was dramatically reduced, over only five centuries, from two thousand to three hundred, and, in consequence, became much more generally used.

Even though the proportion of the population able to use the new, simpler pictographs was still less than one percent, it was now possible to organize a much more complex social structure and to institute the beginnings of a properly bureaucratic cutting-up of social activity. Several cities joined in federations finally to be ruled by a single king. With this arrangement came the regularization of a wider view of the "place" as a complex, rigidly layered, hierarchical entity, which we in the modern world describe as the "state." The apparatus of community organization was now established in at least three classes: lowest-ranking administrators (overseers and foremen, supervising laborers and farmers), the overseers of the overseers (working in offices, running schedules), and, at the highest level, the policymakers.

This division of labor was made possible by the technology of writing and, for the same reason, so too was conformity in public thought and behavior on a scale not possible before. In simpler times, orally transmitted rituals and orders might, however well-remembered and rehearsed, have become deliberately or unconsciously distorted, but now they could be made permanent and codified in writing. This left little room for the deft avoidance of duty. Axemaker writing unified the Mesopotamian command-and-control system through the bureaucracy, and by the third millennium this authority began to extend beyond work-oriented organization into the behavior of individuals in their private lives.

What happened next was a phenomenon that singles out Mesopotamia among the other great riverine civilizations of the period in India, China, and Egypt. In Mesopotamia, the extension of social control through the use of literacy radically altered the relationship between individuals and between them and the city authority, thanks to the invention of law.

One of the earliest examples of the new rule of law was the way it took redress of grievance out of the hands of a victim's kin. The way it did so was special to Mesopotamia. Earlier, in hunter-gatherer times, it seems that wrong-doing had been punished through acts of revenge by family members, in which the injured party or their kin acted as both judge and executioner. The problem with this system was that revenge could turn the original wrongdoer into an injured party, whose kin might then in turn desire their own revenge. This process could all too easily give rise to blood feuds lasting for generations. This kind of "loose cannon" behavior within the close confines of city walls could easily damage the cohesion of the community. The creation, by axemaker tools, of a "place," in the form of these first tightly organized cities, had now begun to affect the values and ethics of those living a highly regimented life within their walls.

In India, China, and Egypt, as religion became institutionalized, the responsibility for judgment and punishment seems to have been transferred from heads of families to the overall authority of the priesthood. Antisocial acts were made even more undesirable by being defined as an affront to the gods.

However, Mesopotamia broke the pattern, thanks to an early shift from communal to private ownership of property as far back as 4,500 years ago, when Mesopotamian clay tablets were already carrying details of private transactions and contractual agreements. Perhaps this kind of low-level commercial interaction needed something simpler than the full ceremony of reference upwards to the gods, and for that reason a lower-level form of regulatory authority came into being.

This, the first written law, differed from other institutions set up so far for the purpose of social control. Since it dealt primarily with private property, it also had to take into account a radically new concept. The axemakers had provided a gift that, throughout history from then on, would color all social development—the idea of the individual's rights and duties of ownership. For the first time, a template for behavior existed that could be adapted to any individual circumstance. Individuals' lives were no longer to be lived entirely at the whim of priest or king. On the other hand, those lives were no longer entirely their own, thanks to the development of law.

All early Mesopotamian law is little more than a collection of precedents, key among which are the references to the special form of the ruler's authority. This authorized the intrusion of power into the private lives of each individual on pain of heavenly retribution. All edicts begin with a statement of the king's appointment by the gods to rule over the city-state, then follows the body of law, and the edict ends with a warning that anybody defying the king's law will be cursed.

Four thousand years ago, in what may be a reference to the first attempt at rule by law, King Ur-Engur of the city of Ur says he administered justice "following the laws of the gods." This gave added magic force to the new code for behavior by suggesting that any transgressor who disobeyed the king would be punished by authorities as far away as heaven. You could run, but you couldn't hide.

The oldest known formal legal code is that of King Ur-Nammu of the city of Nippur, founder of the third Ur dynasty, 4,050 years ago. Among other things, it refers to another major element of social control, reflecting the extent to which the axemaker gift of numbers had developed to become a tool to standardize behavior and regulate human relationships in commerce. This new constraint on thinking came in the form of a standardized system of weights and measures. The world could now be officially packaged. Increasingly, standardized commodities and standardized behavior life was becoming routine.

The relevant court of law imposing order on disputes was, not surprisingly, the temple. Hundreds of tablets unearthed in the ruins of Nippur, in the area where the temple scribes lived, all paint a clear picture of the administration of justice during the period. Court records were known as *ditilla,* meaning a "completed lawsuit," and among them are notarizations of agreements as well as contracts referring to marriages, divorces, child support, gifts, sales, inheritance, slaves, and the hiring of boats, pretrial agreements, subpoenas, thefts, and damage to property. Justice was administered by the *ensi,* the governors of the cities within the city-state, who represented the king. The temple served as the courtroom and there were no professional judges, since the thirty-six men listed as judges were from all walks of life: merchants, scribes, temple administrators, and high officials of some kind.

From the beginning of the third millennium B.C.E., five thousand years ago, groups of *Amurru,* Amorite Semitic invaders from the Syrian and Arabian deserts, made sporadic but increasingly frequent attacks on Mesopotamia. The city-states finally fell, the Sumerians ceased to exist as an ethnic, linguistic, and political entity, and the Amorites established a dynasty that was to rule from its center at Babylon for three hundred years. In 1792 B.C.E. the Babylonian king Hammurabi came to power, and under his supreme authority Babylonian government became highly centralized, with a tight network of governors and officials, who represented the king's interests in all aspects of public life.

Forty-two years later Hammurabi issued the now-famous legal Code of Hammurabi. In its prologue, he claims to have been elected by the gods to reign over Babylon and to preserve justice among its people, ending with blessings for those who respect its laws and curses for those who do not. The Code's epilogue states: "These are the prescriptions for justice established by Hammurabi, in accordance with the wishes of the gods, who guided the kingdom along the proper [*sic*] path." The epilogue reveals an awareness of the strength of legislation as a tool of social reform, designed to prevent oppression and facilitate justice.

The Code was divided into three parts, the central section containing 282 clauses and showing clear evidence of a drive toward more permanent, systematic legislation. Significantly, the Code is carved on a durable stone column, not written on fragile tablets of clay. The provisions of the code are more numerous and much more secular than the codes that preceded it. And for the first time, the "eye for an eye" canon is introduced as a deterrent. The law, for instance, now mandated the death penalty on occasions where, before, it might have been enough to require retribution payment, sanctioned by the temple and due from the wrongdoer's kin.

The Mesopotamian concern for the continuity of social order, ensured by the rigid codification of the individual's rights and responsibilities in law, has informed the whole of subsequent Western thinking, especially in the way it supports the separation of society into classes under the supreme monarch, who rules by divine right. A major example of the way axemaker gifts profoundly modify our thinking lies in the way this gift (that so drastically reduced our ancient, hunter-gatherer liberties) radically altered our view of individual behavior, to an extent that still colors our attitudes thousands of years later. We in the modern world refer to "freedom under law" when we speak of what our ancient ancestors would surely have regarded as major constraints on all their freedoms.

But the modification of thinking brought about by writing and commerce pales into insignificance beside what happened because of the interaction between them. Trade in each of the riverine civilizations was facilitated immensely by the esoteric ability to write, and it was in Egypt that key advances in writing technology occurred, thanks to the plentiful supply of a much more versatile and portable writing medium than the clay tablets of Mesopotamia. Although the presence of the state authority was always visible in the form of hieroglyphs on virtually every structure, like Soviet factory slogans, reminding the people of the plans and achievements of omnipotent Pharaoh, the Egyptian economy was driven by a plant. The papyrus, growing in profusion along the Nile, made possible the easy spread of literacy among bureaucrats because its leaves were prepared simply and written on with brush and ink.

Two versions of writing developed, one called "hieratic," for use in religious writing and official statements, and the other a simpler, popular form, called "demotic," that dealt with abstract concepts. With a highly flexible form of communication like this, the Egyptian elite soon built an empire unrivaled in the Mediterranean. Egyptians traded with China through Indian middlemen; they went to the Atlantic and traveled as far south as central Africa.

Egypt developed slightly later than Mesopotamia and, given her differing environmental circumstances, took an alternate route to civilization. To begin with, the hunter-gatherers in Egypt had long been protected by natural barriers against enemy incursion: to the north lay the sea, and in all other directions, the desert. There was no need, as there had been in Mesopotamia, to develop separate, independent village-cities, each self-protected from attack. The flooding of the Nile was extremely regular, making the early development of large-scale, centralized public-works irrigation schemes relatively easy.

From the beginning, in Egypt, the common experience of a great life-giving river may have originally generated common myths and beliefs that were shared by tribal communities separated by distance but united by the Nile and making their later integration easier. Given the homogeneous nature of the community, it was relatively easy for a single, supreme authority to appear almost from the start, and by 5,000 years ago the centralization of authority in Egypt was total.

Pharaoh's administrators were members of his family, he ruled alone by divine right, and there was no legal system like that in Mesopotamia, since there is no mention of merchants and the concept of private property that had generated Mesopotamian legal codes. Egyptian bureaucracy was closely associated with the temple-court, and it controlled all trade in and out of the country. Since everything in Egypt literally belonged to the Pharaoh, the distributor of all benefits, he alone issued governmental edicts regarding everything from regulations for timber cutting to irrigation, shipbuilding, agricultural practices, and commercial travel.

Egypt attained the most bureaucratized economy in history, thanks to the rigid stratification of its community. The extreme, centrally organized division of labor generated an economy that consisted basically of specialized, exclusive crafts and massive state-controlled labor projects, like irrigation canal networks, sacred cities of the dead, and pyramids. What the Egyptians accomplished was not very different or innovative, but it was uncommonly big.

The dynastic Egyptian state marked a new stage in the expression of power. The institutionalization of control through technology and writing enshrined preferential treatment for the literate. The gulf between the aristocratic elite and the powerless, passive commoners was sanctioned by practice and ritual. In light of this, it is not surprising that here, as in all ancient communities, the first law to be formalized by the central authority regarded acts of *lèse majesté.*

Almost all of the technological advance was pressed into service to help government to function and to control the community. Writing and mathematics serviced taxation and organization. Specialized metallurgical crafts created weapons of conquest and luxury objects of veneration. Calendrical knowledge, astronomy, and geometry were developed specifically for state irrigation projects or to invest the authorities with the magic power of eclipse prediction.

And then, approximately 3,600 years ago, came a major development that would make the acquisition and application of knowledge immensely easier and that would once again alter the way the Western brain worked. Its emergence would also signal the beginning of the end of our millennial reliance on tradition, ritual, and divine authority. The new axemaker product would free rulers from any limitations on their freedom of action that oral tradition might have imposed, because it made the management of higher rates of innovation and change much easier. It was a new kind of script, one that was the first truly transparent communications system because it could be used to express any and all languages. It was the alphabet.

It first appears in one of Egypt's overseas enterprises, a turquoise mine in the mountains of the southern Sinai peninsula, at a place now called Serabit el Khadem. The building complex there included a temple to Hathor, goddess of turquoise, and a large building, including a courtyard, sanctuaries, baths, and soldiers' barracks. The mining personnel were Semitic slave laborers and the enterprise was managed by Canaanites, who spoke a Semitic language close to ancient Hebrew.

These Canaanite mining specialists had been trained at an Egyptian mercantile center and would have been aware of the main commercial forms of writing at the time—hieroglyphs and pictographs—neither of which suited the Canaanite language and both of which were still difficult and complex to write. It might have been the search for an easier way to do things that drove one of the Canaanites to think up a simpler form of expression. Or perhaps the invention had happened elsewhere among the Semites and was brought to Sinai by the miners. But whoever did it and for whatever purpose the new tool was invented it must have made intercommunity trade and technological advance easier almost at a stroke.

The Egyptians (and the Mesopotamians, Cretans, Cypriotes, and Western Semitic peoples) had already begun to abbreviate their complex forms of picture-writing by using syllabaries. A syllabary reduced the number of signs by adopting a common sign for all examples of a single consonant. This made much easier the use of Egyptian hieroglyphs, for instance, that included no fewer than 700 signs. The syllabary was achieved by taking the sign for a word with that consonant's sound and making it the sign for the sound wherever it appeared (e.g., "mayem," the wavy sign for "water," was used to represent the letter "M" because the spoken word had an "M" sound).

However, what made the syllabary difficult to use easily in other languages (unlike the alphabet) was the way that, in Egyptian, for instance, twenty-four signs were used to express all the possible modifications of a consonant by a vowel ("ma," "mo," "mi," etc.), and a further eighty signs represented a pair of consonants modified by *two* vowels (e.g., "timi," "tama," "tima," etc.). And since some of the vocalic sounds involved in syllabaries were not necessarily simple vowels, or they were present only in one particular language, in a kind of linguistic catch-22, it was necessary for a reader to know the language in order to reproduce the sound.

The Sinai scribe probably developed his technique from a syllabary known to him (perhaps Western Semitic, used in Phoenician, Hebrew, and Aramaic), by simplifying and reducing the number of letters needed. All he did was remove the modified forms. In this way, a written transcription of the sound system could satisfy the needs of any language, without having to refer to the full pictographic signs that went together to make up the word.

The signs found at Serabit el Khadem, scratched on the limestone rock faces, are of letters written in a flowing style, indicating an origin in brush and ink rather than in carved stone. The unknown Canaanite axemaker had invented the first true alphabet in order to make it easier to do business deals between members of different linguistic groups, but when the new script eventually reached Greece, it would do much more: it would trigger the beginning of modern thought.

So in the near ten-thousand-year period, from the first agricultural settlements to the creation of numerals and the alphabet community, hierarchs had used the axemaker gifts to maintain, strengthen, and centralize their grip on society while at the same time providing an increasing number of its members with the means to lead fuller and more satisfying material lives. But the gulf was widening all the time between the few who understood the esoteric knowledge that conferred the power of social cut-and-control and the many who did not. And in spite of the fact that, from the shaman's baton to the alphabet, toolmaking generated an ever-growing amount of ever more accessible knowledge, it must be remembered that at no time was this access ever available to more than a tiny fraction of the population.

And as knowledge expanded, so did specialization and esoteric practices. Above all, increased knowledge created more complex societies and activities, and these demanded ever more careful management. The consequences of social breakdown, in the crowded confines of a Mesopotamian city reliant on organizational conformity to ensure the continued provision of food, were potentially much more dangerous than they would have been among a small, free-ranging, hunter-gatherer group ten thousand years earlier. Besides, the prospect of security and continuity must have been attractive not only to the few who would profit immensely from a concentration of labor inside city walls, but also to the great unwashed, living on the precarious line between feast and famine, hemmed in by the same walls and cut off from the source of their food and clothing. Conformity and obedience in these circumstances was pragmatic. The devil, you say.

Even by this early date, axemaker gifts had already given us the ability to perform miracles. We had used them to emerge from the jungle, first to small, regularly fed agricultural settlements and then to large, well-ordered cities. There, in return for the security of protection, possessions, and food, we traded the ancient hunter-gatherer freedom of movement and the right to change our leaders, for royal dynasties that ruled by divine right and codified our behavior with the rule of law.

Concentrated and regimented in the cities, bound by rigid conformity, we were conveniently ready for the next great axemaker change. In return for the gift of the alphabet we had accepted, we would have to accept a new measure of conformity in the way we thought about thinking.

IF
A
=
B
AND
TO: THE GREEKS
FROM: PHOENICIANS

Chapter 3

THE ABC OF LOGIC

I invented for them the art of numbering, the basis of all sciences, and the art of combining letters, memory of all things, mother of the Muses, and source of all the arts.

AESCHYLUS, *PROMETHEUS BOUND*

Axemaker gifts often trigger self-fulfilling prophecies because they create problems that only they can solve. In Egypt and Mesopotamia and the other riverine civilizations, the fact of living together in such large numbers (made inevitable by the reasons for which we had been led to settle down in the first place) created the need to organize and quantify the products of the agricultural techniques we then used to survive. The food surplus raised the population and boosted commerce to the point where regulation through writing was the only alternative to chaos. Regulation in turn standardized behavior through legal regimentation. And, of necessity, living within walls brought a new hierarchical attitude toward each other. We had been freed from the vagaries of nature to be deviled by regular meals.

Now the same kind of cycle was about to run again. A few centuries after its invention, the alphabet would be picked up by the Greeks, a group of people with a particularly pragmatic lifestyle, many of them seafarers with the free-wheeling curiosity such people often develop, dealing as they do with frequently changing circumstances, not least the weather and unknown landfalls. The alphabet would eventually offer them (and then us) a means to satisfy their curiosity to a massively greater extent, but it would also limit them (and then us) to a new kind of constraint: alphabetic thinking.

That constraint is operating as you read these words now. You are so completely reprogrammed by two and a half thousand years of the alphabetic process that as you read this text it seems perfectly normal that the individual letters you are reading merge into words. And they "naturally" arrange themselves in a straight line as your eye moves from left to right. Yet these letters themselves have existed only for a short time and briefer yet has been the time they have been read from left to right.

Human beings have recorded information in very many different forms and formats: signs, tallies, numbers, or shorthand. The signs can be in cursive script, written top to bottom or bottom to top. Or the direction of reading can alternate first left, then back to the right, or first up and then down. It can be expressed in pictograms, read up and down, right to left, or in a spiral, or even in a brick shape, and lots more.

One format, our alphabet of twenty-six letters, written from left to right, achieved its modern form in Greece 2,500 years ago. And like the shaman's carved baton, the moveable type of Gutenberg, and the electronic computer, it was one of the major building blocks of modern thought.

After the first primitive examples at Serabit el Khadem, the next appearance of the alphabet occurs in Phoenicia, modern Lebanon. The first full text consists of no more than a few words, written using a twenty-two consonant-letter system, on the sarcophagus of the Phoenician king Ahiram of Byblos in 1,000 B.C.E. The new alphabet must have been irresistible to the Phoenicians because, as has been said, the phonetic basis for the letters made communication with any language easier. The Phoenicians were in constant touch with foreign communities because they were the greatest travelers in the ancient Mediterranean.

They appear in Homer: "famous as seamen and tricksters, bringing tens of thousands of trinkets in their black ships." Loaded with merchandise, Phoenician merchantmen plied their trade to every corner of the known world. Phoenicia exported pine and cedar wood from Lebanon, fine linens from Byblos and Tyre, metal and glass, salt, and fish. They imported precious metals and gems, papyrus, ostrich eggs, ivory, silk, spices, and horses. They discovered the dye made from a shellfish so rare and the product so costly it has been known ever since as "royal purple." These wide-ranging Phoenician traders took their new alphabet with them, and their inscriptions have been found as far afield as Cyprus, Marseilles, Spain, Sardinia, and Malta.

Then, at some time in the ninth century B.C.E., the Phoenicians came across the first Greek colonies, north of them on the Asia Minor (Turkish) coast, on the island of Rhodes, and perhaps also in Crete and Cyprus. The encounter was to be highly significant for the Western world. The meeting is indicated by a few archeological finds of eighth-century Greek alphabetic inscriptions, mainly on a number of pots and, more widely, a century later, in the names of the dead inscribed on monuments, graves, or dedications to deities. Sometimes there is also an alphabetic version of a potter's name. The Greeks took over the alphabetic signs and simply kept the Phoenician names for them, referring to their new letters as *phoenikia,* meaning "Phoenician things."

At the time of this contact, the Greeks had only recently recovered from the centuries-long period of chaos, following the fall of the city-state of Mycenae and were expanding from their mainland, out across the Aegean. By now they boasted a sophisticated social system, traditional laws, and a body of mythical, oral knowledge handed down in the form of epic poetry sung by bards. They had already indicated their eagerness to innovate by taking over arithmetic from Mesopotamia, simple geometry from Egypt, and metallurgy from the Assyrians. Greek colonial city-states like Miletus, on what is now the Turkish coast, were already rich, their dynamic economies based on seaborne trade with other coastal communities.

The Greeks seem to have encouraged small groups of Phoenicians to settle and trade with them so that they could learn Phoenician techniques for making jewelry, and cosmetics. Where the actual alphabet transfer took place remains a mystery. One site might have been an eighth-century Greek trading settlement on the Syrian coast, at what is now El Mina. Crete and Rhodes also provide evidence of contact, with Phoenician and Syrian imports appearing in both places in the ninth century B.C.E. There were, for instance, Semitic jewelers living in Rhodes from around 1,000 B.C.E. An inscribed Phoenician bowl dating from the ninth century B.C.E. has been found in Cyprus. Semitic luxury goods were also reaching the Greek mainland at the coastal island of Euboea around the same time.

But wherever the momentous event took place, a single location is likely, because in every version of the new alphabet found afterwards throughout Greece, the same mix-up was made in transcribing the same four Phoenician sounds: "zayin," "tsade," "samekh," and "s(h)in." The sign for zayin mistakenly became that for tsade (Gk.: zeta), for tsade that of zayin (Gk.: san), for samekh that of s(h)in (Gk.: xi), and for s(h)in that of samekh (Gk.: sigma).

The Greeks also altered the alphabet slightly, because their language used fewer sibilants. Not long after the transfer, they also added letters for sounds specific to Greek: phi, chi, psi, and omega. This modified Greek version of *phoenikia* was eventually able to represent all the sounds of speech in such a way that they could be easily read back. In other words, it solved the age-old problem of how to produce shapes that automatically triggered an acoustic memory.

The new alphabet was, in fact, seen first as an aid to memory, in an oral or preliterate culture where memory played a far larger part than it does in ours. In these cultures, the traditions of the society are transmitted by storytellers. This means that the rituals, rules, manners, and history of the society have to be remembered by somebody. And as the society gets more complex, of course this ability is taxed to the limit.

However, it would be a mistake to regard Greek society before the development of the alphabet as primitive or unsophisticated. Architecture and geometry were well developed, and Homer's magnificent *Odyssey* and *Iliad* almost certainly date from this time, as do some of the very early philosophers. Education was oral, and it concentrated on music, memorizing poetry, reciting, and singing. The function of poetry was used not to express emotion of imaginative thoughts, but as a means of keeping record. Memorizing necessary facts was easier if there were a trick to it, such as rhyme. We see the same technique being used in late Middle Ages England with doggerel like:

> Red sky at night, shepherd's delight.
> Red sky at morning, shepherd's warning.

One of the first uses of the alphabet was to transcribe this kind of oral tradition, which is why early Greek literature was a kind of poetic data sheet. It is impossible to understand fully if it is thought of as literature, in the modern sense. In Classical Greek, writing was initially used as a transition between oral and documented record. We moderns, living in a literate society, see written prose as prosaic and poetry as poetic, whereas at the beginning in Greece this position was reversed. And because we regard reading and writing as so important, we tend to assume that the first people to have acquired it would have been the aristocracy. In fact, they were among the last to become literate.

Although the Greeks may initially have seen writing as little more than an *aide-mémoire,* it soon began to do more and eventually changed the way people thought. First of all, the alphabet converted traditional knowledge into an external object easily available for inspection, no longer dependent on memory. The result of this was that new ways of talking and thinking about the world became possible. For example, the oral tradition works very well when describing action, as in the many battle scenes in Homer and later in the Anglo-Saxon poems such as *Beowulf.* On the other hand, a written culture, in which words and ideas can be studied at leisure, is much better at reflective pursuits, such as philosophy, narrative prose, and the like.

Key to the new way of thinking is that where pictographs were in a sense representations of the object, alphabetic letters were not. For example, an "A" represented nothing specific in nature. The alphabet codified nature into something abstract, to be cut and controlled impersonally. In this way, to some extent the alphabet removed us one more step away from our environment. It also gave us a new view of the past.

While literature was one obvious offspring of the development of the alphabet, a less obvious outcome was the concept of history. Oral memory deals with the present, and recollection is concerned with what is relevant to the present. Biography in an oral tradition is not as much careful scholarship as it is a creative act, in which events are woven into coherence with the aid of imagination. But the accumulation of written records makes it possible to separate the present from the past. Somebody who can read is able to "look back" at what happened before, in a way that the nonliterate person never can. Written material is by necessity "dated" and fixed, while an oral tradition is "living" and fluid. In this sense, Herodotus was not so much the Father of History as a child of the alphabet.

Unlike the earliest writing in Mesopotamia and Egypt, the first Greek alphabet was not used for administration or lists of accounts. This is surprising, given that the alphabet transfer occurred in a trading settlement. In fact, the first use (around 800 B.C.E., in Crete) occurred in public announcements of a series of Greek and non-Greek laws carved in alphabet on a temple wall. The new alphabet also appeared on personal luxury articles such as pottery, where the owner's name was often included. On the Italian island of Ischia, in 720 B.C.E. a vase known as "Nestor's Cup" carried an alphabetic message that read: "Nestor's cup was fine to drink from. But whomever drinks from this will be taken by desire for Aphrodite."

The alphabet caused an immediate revolution in the way society was structured because it was so easy to learn, so that many people could now read. Many cultures have managed well with pictographic and other writing systems, but the modern expansion of literacy and democracy came primarily because of the simplicity of the Greek alphabet. Even today, because of the complexity of their writing system, a culture such as that of the Japanese, which may well have adopted many elements of Western life, still requires an education that takes years longer (and places students under more pressure) than does that of the West. But back to Greece.

With the Greek alphabet, humans had for the first time an easy-to-use "external data storage system" that compensated for the considerable limitations affecting human memory. Immediate memory only survives for two seconds without rehearsal, memory for lists of words lasts only about fifteen seconds, and humans can only take about five to seven items into their temporary store. Above all, the whole system is very vulnerable to interference, as can be seen in eyewitness memory, which is notoriously unreliable.

However, with the aid of writing, the brain can manipulate symbols and ideas without having to expend the effort necessary to reproduce them. In modern cultures, people who are engaged in abstract thinking use external material, like writing, as their "working memory." External storage also *publicizes* thinking, so that ideas can be considered, commented on, and criticized. Science is perhaps the most powerful example of what this ability can make possible. So for these reasons, the "literates" in the Greek community now had a tool to chop up thought and ask complex questions without having to worry about getting lost in the process.

This doesn't mean, of course, that particular individuals in earlier communities, like the Hebrews, weren't capable of abstract thought and reasoning, but those abilities had been confined to an extremely small elite. Although the alphabet did make knowledge more accessible, literacy still wasn't for everybody. Only the people in positions of power got to read and write.

In spite of this, though, the introduction of Greek letters and the wider literacy it generated was to alter the character of human culture and separate alphabetic societies from their oral contemporaries. It made possible a democratic form of government and a faster, more effective educational system. It was now no longer necessary for children to memorize hundreds of icons, or to regurgitate community knowledge in difficult, time-wasting poetic recitations, as had been the case in Greece for over a thousand years. Perhaps most important of all, the alphabet was another axemaker gift that would change the way the human brain actually functioned and because of that the way alphabetic humans saw themselves and their relationship to the world.

As said, written language can take many forms: down-up, up-down, right-upper to left lower, down first to the end of the page or space and then up, or right-to-left to the line-end, then returning left-to-right (in Greek called *boustrophedon,* after the back-and-forth route an ox-drawn plough takes over a field). Writing can also radiate out from the center, or form a spiral. In contrast, ancient hieroglyphs tended to go only from right to left.

An analysis of the world's writing arrangements by Derrick DeKerckhove discloses that all systems that represent sounds are written horizontally and all that represent images, like the Chinese, are represented vertically. And, more to the point, all systems that contain vowel sounds are written (except for Etruscan) right to left.

It is probably not a complete accident that the Greek alphabet became left-to-right soon after the acquisition of vowels, given the increased ease with which the left hemisphere of the brain could process script read in this way. It probably happened as some individuals found superiority using only half the *boustrophedon* movements. Around this time (500 B.C.E.) many revolutions in thought besides the Greek were emerging around the world, including that of Confucius in China and of the Buddha in India. However, the alphabet made a special contribution to the human ability to dissect and reshape the world. This development may well have provided the necessary final set of components which would be the foundation of our "modern" way of thinking, which we, of course, identify as having begun with the Greeks.

Left-to-right orthography is read differently. The eye movement control, moving to the right, is run by the left hemisphere of the brain. So left-to-right letters are first seen in the right-hand field-of-view of each eye and then processed in the left hemisphere of the brain, which is specialized for sequential, bit-by-bit processing and the analysis of chunks of information.

So writing this way encourages the processing of language, as if a word were the product of serially assembled, cut-up pieces, rather than an illustration. And it is possible, with left-to-right "en-voweled" writing, to understand a text even if you have no previous knowledge of the subject (as was obligatory with earlier, iconic nature forms of writing, where symbols represented things rather than sounds), or even the word, simply by reproducing the sequential sounds and building up the word bit by bit.

Although this is a very speculative argument, children growing up in that newly alphabetic world, like those growing up earlier in a newly carpentered world, may well have experienced different brain development. The reason may be that the easy to pick up Greek alphabet allowed children to learn to read while they were still learning their oral vocabulary. Thus reading and writing could be taught much more easily during the period of a child's growth, when its brain was still developing linguistic competence. With left-to-right reading, the ability to represent the world abstractly and to combine and recombine abstract elements then became part of the way we were educated to think about the world.

All these components, working together with a developed social order, established food supplies, and community security combined to make possible the first leap to modern conscious knowledge, because there was now a literate culture in which there could be a distance between the thinker and the thought, through an externalization not only of the memory (as with the ancient batons), but of the process of thinking itself. This way of treating knowledge as an artifact would now further separate axemakers from the rest, because it would make knowledge a new world in itself, to be cut up and segmented by specialists.

This change in thought processes can be seen within a century in the rise to prominence of a way of looking at the world analytically, step-by-step, with the development of new Greek procedures for acquiring and analyzing knowledge (which they called "love of wisdom," or *philo-sophia*).

Perhaps because they had this new ability externally to cut and control thought, early Greek intellectuals were largely free of the religious awe that had permeated most thinking until then. Philosophy, as we now know it, had begun to develop a hundred years or so earlier, and for the first time in recorded history, questions were being asked about the nature of knowledge itself, about the practical aspects of the rule of law, and about the establishment of social conventions.

Amongst the first to tackle such issues were the sixth-century-B.C.E. thinkers in Miletus. All of them were practical men involved in the politics and commerce of their city, knowledgeable about math and the geometry. They were the first to give purely natural explanations of the origin of the world, free of mythological ingredients. In general, they tended to make large generalizations on the basis of restricted but carefully checked observations.

The alphabet helped them get away from the old, polytheistic way of thinking and to produce rational and general laws based on sequential cause-and-effect explanations for natural phenomena. For this reason, they asked new *kinds* of questions that related the particular ("Why does fire smelt metal?") to the general ("What is the nature of fire?"). This new process set the pattern for the ways humans have thought about the world ever since, and for many issues that are still central to modern thinking. For example, one of the most fundamental divisions in the way humans think is between the critical, analytical style that tries to understand the world by examining its parts, and the synthesizing, speculative style that attempts to describe the whole.

The Milesian philosopher-axemakers thought about abstract topics that still exercise our minds today: is there one basic element in nature that combines to make everything, or are there several different elements that mix together to make the world? Parmenides decided there was only one enduring nature, whereas Heracleitus claimed that all was a constant, changing flux. Another question they asked was whether the basic substance of the world was continuous and fluid or discrete and atomic? We now think it is atomic, just like Democritus of Abdera, who, taking as his metaphor the way words were made up bit-by-bit with the alphabet, taught that the material world is made up of particles that emerge, combine, and recombine from a primeval slime.

The alphabetic process of making words by taking a set of abstract elements and recombining them in myriad forms accelerated the Greek view that this was also the way the material world worked. As letters composed many words, so atoms with different shapes and sizes might possibly compose many things. In this way, different substances might also have different properties because their atoms were differently shaped, differently placed, and differently grouped.

Aristotle referred to this view in his *Metaphysics:*

> And just as those thinkers [Leucippus and Democritus], who posit one underlying substance, generate all other things by its *attributes,* positing the *Rare* and the *Dense* as the principles of all other changing attributes, so these thinkers say that the differentiae are the causes of all other differences in things. These differentiae are three: *Shape, Order,* and *Position.* For they say that things differ only in contour, arrangement, and turning; and of these, contour is shape, arrangement is order, and turning is position. For A differs from N in shape, AN from NA in order, and Z from N in position.

Since so many people could now read, this made possible a debate among the literate members of the culture. Aristotle, for instance, pointed out the problems the atomists face when trying to explain changes in physical states, such as what happens when things melt or evaporate. The Greeks also wrestled with the question that still bedevils Big Bang believers today: how does something come out of nothing? One of the Milesians, Anaximander, about 150 years before Plato, suggested that the world developed out of *apeiron,* which was infinite and indefinite and out of which emerged the four basic elements: hot and cold, wet and dry. His disciple Anaximenes took this further and suggested that the original matter of the world remains the same through all its transmutations (not a bad early stab at the law of thermodynamics).

There was even a crude forerunner of evolutionary theory and the survival of the fittest by Empedocles, who said that in the beginning there had been individual parts of animals moving about (heads, arms, legs) which then combined into all sorts of monsters, of which only the most organized survived. Anaximander proposed that land animals were originally sea animals and that humans were descended from the sea animals that carry their young in a pouch.

It is not just that *some* Greek insights have held up so well (since many, of course, haven't), but that their style of thought and the debates that raged among them broke with the type of thinking that had held sway up to then. Instead of magical, superstitious thinking that assigned causes on the basis of similarities or attributed unlimited and arbitrary power to the gods, theirs was an attempt to construct an explanation for the world in terms of what could be observed according to regular principles. In other words, with alphabetic thinking the world could be mentally held at arm's length, examined, and discussed.

The dissections and analyses made possible by the alphabet distanced the illiterate person in the street more than ever from an understanding of the world and how it functions. In typical axe-maker style, the new thinkers described a "real" world very different from what appeared to the untrained perceiver. Although the eye might see trees and houses and people, what was really there were atoms, an everlasting underlying substance or, as Pythagoras thought, only numbers.

From the time of the first axe, tools had changed the world and the alphabet was the most powerful tool for change so far. But it was also a tool that was so easy to use. If literacy spread too wide, the result might mean loss of control for those in power. This danger would first be revealed by a mundane problem caused by the expansion and differentiation of the Greek economy to the point where there was an urgent need for a larger number of educated citizens. Since the sixth century B.C.E., primary education had been generally available for all free men, but the system was not well developed enough to provide a trained cadre of bureaucrats who could handle the founding of colonies, run the public treasury, finance and organize war, collect customs tax, and so on.

Administrators needed to know about geography, economics, constitutional law, foreign customs, and a lot more besides. Above all, clear, skilled speech was essential for the government of a complex but orderly society in the new Assembly, where polity was public speaking and government functioned through persuasion by argument. This ability to argue and debate was highly developed by a group called the Sophists, who took what the alphabet could do to an inevitable conclusion.

The Sophist Protagoras was the first to suggest the potential value of the way in which sequential and combinatory thought made it possible to argue two sides of any issue. This introduced the socially divisive possibility that there might be no absolute truths or morals, since these differed from one community to another and, in any case, were, by definition, now "arguable."

It is this relativism, the very thinking that made them outcasts in their own society, that attracts us most to the Sophists. At the heart of their approach was the idea that (like Democritus' atoms) concepts and ideas were changeable elements just like the letters of the alphabet, and so moral values, for instance, were relative. What was thought to be good and valued in one society was often scorned in another.

Even more sophisticated was their idea that all knowledge was equally relative and that what counted as knowledge was not an absolute, but a view of the world strongly influenced by the society of the time and place. This was an offshoot of their thought that man and his values were at the center of any interpretation of the universe.

Another Sophist, Georgias of Leontini, was born shortly before 480 B.C.E. in Leontini, in what is now Sicily, and lived there for much of his life, although he visited Athens as an ambassador to ask for aid. As a tribute to his oratorical powers, a golden statue, that he dedicated himself, was erected in his honor at Delphi. He never married, left no children, and died at the court of Jason at Pherae, in Thessaly, aged between 105 and 109 (sources differ). When asked the secret of his old age, he said that it was because he never attended other people's parties.

Georgias placed into the Greek theater of ideas some of the fundamental issues in philosophy, with which we still grapple today. His subject matter has an unusually modern ring. One of the Sophists' special skills was rhetoric, the art of presenting an argument so as to convince the listener. Georgias invented a lecturing style that involved conducting his lectures in the form of a debate. He would take first one side, then the other, and then give a supporting speech for either side, emphasizing the arbitrary, cut-and-combine nature of language. Plato complained that Georgias' speeches could make "small things seem large and large things seem small by some power of language and new things seem old fashioned and vice versa."

But this emphasis that Georgias and the other Sophists placed on rhetoric was not just related to swaying political opinion. It came from a realization that the relationship between speech and "truth" is far from simple. Speech is not just a matter of presenting the facts, since considerable reorganization of the "facts" is involved in the way they are selected and sequenced. It was this difference between rhetoric and reality that lead Plato to contrast rhetoric with philosophy and to condemn it.

The effect of the new literacy, the power of the word, and the difference between words and reality was something Georgias explored further in his most famous work, or rather, the bit of it that survives: *On the Non-existence of Existence, or On Nature.* This made Georgias infamous in his day and caused one contemporary to remark: "How could one out-do Georgias, who dared to say that, of existing things, none exist?"

In the book Georgias argued that:

1. nothing exists and
2. if it does exist, man can have no knowledge of it, but that
3. if somebody did know of it, he could not communicate it.

The Sophist point was that the elements of thinking remain much the same, whether they are being used to support or attack a rationalist position. This was an arguable view, but the extraordinary thing is that someone should have been playing with ideas in this cut-and-combine manner as early 450 B.C.E.

Georgias was cutting up the concepts "being," "thinking," and "saying," which had previously been unified (and which have to be so, for truth or knowledge to have any meaning). He was raising, so long ago, the question of meaning and reference. If words referred to things, how could they each be connected to each other, in the sense of why a word is attached to one thing and not to another, since there is nothing about a word that links it to its reference?

Georgias held that when we communicate we never exchange the thing but only the word for it, which is always other than the thing itself. So every word introduces falsification of the thing it refers to, and this means that one can never reproduce reality and that any claim to be able to do so is a deception. But since this is exactly what all words claim, then all words are deceptions. If this is so, then the man who communicates best deceives most. While in the modern world this thought has a faintly political ring to it, the ancient Greeks lived in days before television.

But the Sophists' moral relativism still provokes today's conservative thinkers, especially in regard to the Sophist view that there were no absolutes when it came to human behavior. In regard to sex or marriage or burial, for instance, what people did depended on the culture they lived in, so the behavioral rules regarding these events were just conventions. The important thing, the Sophists concluded (foreshadowing our modern armies of advertising executives, marketing people, and spin doctors), was to gain influence over others, and this was what they promised to teach people (for a fee). Their outlook became most clear in the teachings of Thrasymachus of Chalcedon who declared: "Right is what is beneficial for the stronger."

While we've highlighted Georgias, a little-known figure in Greek thought, to emphasize the depth of the cognitive revolution happening at the time, his major contemporary thinkers were no less impressive. Of course, although we're not discussing him at length, towering above all these figures in the extent of his influence is Plato, who developed a unified theory involving most branches of learning, such as politics, law, and the arts, as well as the nature of the world. A major influence on his political thought was the disaster of Peloponnesian Wars, in which the Athenians had been defeated by Sparta. So Plato swung between rejecting democracy because it could lead to the kind of demagoguery that had driven Athens into that war and at the same time being horrified by excesses of the subsequent dictatorship. He developed one of the first proposals for how a country should ideally be governed. In this system, in return for absolute power enforced by Warriors, Philosopher-Guardians would act selflessly in the interests of the state because they themselves owned no possessions.

But he also saw that any philosophical theory had to explain the natural world. His suggestion about the basic nature of the world was that only the truths of mathematics endure so that only a theory based on numbers with a geometrical framework would reveal the permanent structure behind the obvious change and decay of the world.

He was the archetypal theoretician who believed that the structure of matter could be worked out from logical principles and so there was no need for observation. This approach can be most clearly seen in his idea that heavenly bodies must move in circles because in doing so they were following the most perfect shape. He wasn't interested in the mechanics of how it was done, and the only purpose of observing the stars, he believed, was to find examples of those circular movements. Similar rational principles would reveal the origins of the world. Because the most perfect shapes are mathematical, the world must have begun with two sorts of right-angled triangles, which in turn generated regular solids, which in turn made up the particles of the four elements.

Although Plato opposed the Sophists' emphasis on rhetoric alone (he even wrote a dialogue about Georgias, of which Georgias said: "How well Plato knows how to satirize!"), he produced a monumental series of dialogues. These portray almost every conceivable intellectual position on many issues. And although a number of scholars have interpreted Plato's dialogues as taking one or the other of these positions, he was most likely exposing the weak links in thought in order to help people to appreciate the direction of their thinking and so to correct their mistakes.

It was not, however, the considerable and varied output of Greek philosophers that was finally to be the most influential in shaping our thought, but the work of a single student of Plato. He was the son of Philip of Macedon's doctor, and at the age of seventeen or eighteen he entered the Athenian Academy and became an ardent Platonist. On the death of Plato twenty years later, he left the Academy, partly because of anti-Macedonian feeling and partly because of his dislike of the kind of Platonism being taught there, which followed a tendency to reduce philosophy to mathematics. Moving to the coast of Asia Minor, Aristotle began to consider the problem Plato had left unresolved: how the mind (which is separate from and superior to the world) acquires an understanding of matter.

As he moved away from Greece, he moved away from Plato, especially in his views on the observation of nature. Aristotle concentrated especially on biology. In his various works (*History of Animals, On the Parts of Animals,* etc.) he mentions some five hundred different species and appears, in a pioneering effort for the time, to have dissected as many as fifty.

Aristotle's techniques shaped the way you marshal your thoughts and weigh the evidence today when you come to a decision about anything. He formulated the approved rules for how to think and how not to make mistakes. His formula submitted problems to a sequential, rational process. In this way, analysis (according to a clearly defined process) would lead to synthesis through a system of dialogue that made it possible to identify intellectual inconsistency.

This inductive approach involved first recognizing and categorizing contradictory statements. The process of defining the truth was then that of distinguishing between the particular, or local, truth about these statements and the general truth. This was the means by which to arrive at understanding: processing contradictory, incomplete, or inaccurate definitions until a consistent definition emerged.

In an echo of the way the earliest axemaker gifts were used, the enormous value to those in power of this new mental tool was that it let them cut to the core of the world in order to find the essential order in all things and then to use that order to shape social behavior appropriately. The discovery of fundamental order would reveal that the universe was not a haphazard collection of things but had a purpose and design. And the only purpose in life was to understand how society and its individual members fitted into this design.

However much of a democratic or philosophical breakthrough the gift of sequential alphabetic thinking might have been, and whatever options it might have offered for alternative ways of examining the world, deductive reasoning standardized thinking as never before. The new method consisted of propositions taking the form of two premises, each containing a common middle term, with a third, new premise resulting from the other two. The power of this system was that it allowed the thinker to establish the truth about nature even if these truths could not be tested personally or directly. For example: "All that shines in the dark, is fire. Stars shine in the dark. So stars are fires." This method allowed people to see if their mode of thought was consistent.

Aristotle's method was called "logic," and later in history it would be referred to by Islamic thinkers as "the tool for sharpening thought." With it, Aristotle offered a gift of tremendous, world-altering power because it taught a standardized method for chopping up the world, observing it in an orderly way, and analyzing how it worked.

Cut-and-control tools had been used for millennia by those in power to maintain order and stability in their populations, but now Aristotle applied analytic methods to the understanding of the whole of existence. He categorized all living organisms according to a single matrix: the Great Chain of Being. This definition of nature would rule all investigation for the next fifteen hundred years.

Aristotle also constructed a system to explain what happened in the heavens, placing the Earth at the center of the universe, surrounded by concentric invisible spheres each carrying the Sun, the Moon, the planets, and the stars. This universal structure provided a place for everything, so everything was in its proper place. Conformity was all. Like Aristotle's logic, this construct too would rule all thought until Newton.

Aristotle did not believe that it was necessary to suggest a separate realm of ideas to explain how we come to the notion of abstract ideas. He argued that what you see is what you get, that the world of perceived things is the real world, and that there is no great mystery about notions like "good" or "beauty." Having knowledge was to be able to say that something was generally true of certain types, or groups of things. From this starting point a system of knowledge about those things could be constructed. Aristotle really invented scientific method by insisting that observing what goes on in the real world is important. "Credit must be given to theories," he said, "only insofar as they are confirmed by observed facts."

And he practiced what he preached, through simple, systematic observation, noticing all kinds of correlations, such as the fact that no animals have both tusks and horns. He was also interested in embryology and from his observations deduced, for example, that whales were closer to mammals than to fish. In keeping with his respect for the natural world, Aristotle did not accept Plato's idea of a soul existing independently of the body, believing that the soul was the spiritual part of the body and lived and died with it. His followers carried on the approach, identifying hundreds of plant species and discovering how they reproduced (a fact then lost until the eighteenth century).

Around 250 B.C.E., one of the Aristotelians, Strato of Lampsacus, was doing experiments to try and work out what effect fire had by weighing wood before and after heating. It was this Strato who became the head of the great Museion at Alexandria in Egypt, where Archimedes made his famous discoveries about volume and the displacement of bodies in water and where Aristarchus evolved a theory that the Earth rotates daily and moves around the sun in a circular orbit, as do the planets.

State-financed and state-controlled, the Museion was basically an axemaker research-and-teaching institute. It had the largest library in the West, with half a million volumes, as well as rooms for lecture and study, a dissecting theater, an observatory, a zoo, and a botanical garden. A wide range of experiments were carried out, including the investigation of what happened to liquids in a vacuum and how much of the weight of food eaten by a bird was added to its body weight. Many of the hundred or so Museion teachers left their mark on posterity. Alexandrine mathematicians included Euclid, Archimedes, and Apollonius. In medicine, Herophilus established the brain as the center of the nervous system, and Eristratus investigated the cardiovascular system. Eratosthenes calculated the circumference of the Earth, while Aristarchus of Samos advanced the theory that it orbited the Sun. The first systematic grammar was written by a pupil of the Librarian and remained standard for 1,600 years.

But even this apparent spirit of free scientific inquiry was not really free. Everything either fitted Aristotle's models or it was wrong. Artistarchus' heliocentric astronomical model was not widely accepted because it did not match Aristotle's view that the Earth could not follow the same laws as the perfect planets, while the medical teacher and researcher Galen missed the idea of the circulation of the blood because of the similar Aristotelian preconception that on Earth all motion was supposed to be rectilinear. Only in the perfect, heavenly realm could motion be circular.

Aristotle was the supreme axemaker in that he provided a tool for discovery that rulers could unconcernedly allow to be used because it imposed a standard system for manufacturing knowledge within agreed rules. The knowledge generated by this system would provide even more opportunities for social control because, in making possible the analytical cut-and-control of the world, Aristotle opened the door to increased specialization. The gulf between those with knowledge and those without widened again.

The catalyst for change generated by logical, alphabetic thinking can be seen at work in all areas of life, from Greek theater to pottery decoration. In 650 B.C.E., the geometric patterns of the previous generation of potters began to be replaced by representations of human figures when, in the Chigi Vase, naturalistic figures of foot soldiers in phalanx appear. At the same time, in one of the first written texts, Hesoid the farmer-poet, who was trying to classify the gods and myths, argued that social order would only be achieved if humans took responsibility for working out their relationship with the gods, nature, and fellow humans.

The effect of the new, reasoned thought and behavior can be seen in the way the Greeks discussed issues publicly, in the theater. By 450 B.C.E., what had begun centuries before as religious festivals had developed into the Greek tragic form. The tragedies aired the concerns of a society in transition from oral to written modes of thought. They dramatized the social effects of the axemakers' gifts: the changing ways of Greek society, the conflict between the old myths and the new knowledge, the power of the gods contrasted with that of human will.

Perhaps the clearest evidence of this alphabetic effect can be seen in the fifth century B.C.E., when the plays of Aeschylus are joined on stage by those of the newer writers, Sophocles and Euripides. Aeschylan tragedies like the Oresteian trilogy retained the old religious traditions, dealing with the gods and their control over human destiny and written in formal, solemn poetry, with long, elaborate, showpiece choruses. The playwright's concern seemed to be primarily the nature of heavenly government and the way human decisions set in motion unavoidable events foreordained by the gods, who answered disobedience with terrible punishment.

On the other hand, Euripides, a playwright of the new school, wrote in prose and was more concerned with investigating human character. The greatest of his "new-style" plays was probably *Medea,* the story of a woman scorned and wreaking terrible revenge on the man who had betrayed her. The play is the first portrayal of a character that excites pity because of her emotional and psychological torment. It is also perhaps the first play to be written about a human being exercising free will. The gods had no part to play in Medea's story because she herself was unwilling to accept their involvement. With *Medea,* Euripides threw off the old theatrical and religious ritualized convention.

In this new world, individuals were the judges of their own actions. Events no longer followed the dictates of religious ritual but the new (and no less conformist) laws of human society. In *Antigone,* by another of the younger tragedians, Sophocles, the lyric chorus makes the burial of Polyneikes sound like a celebration of the axemakers when it proclaims the ascendancy of humankind over the ancient gods, listing the accomplishments of sea transport, agriculture, hunting, land transport, speech, housing, and medicine. Literate humans were beginning to realize that new knowledge was giving them more power to direct their own destinies and to master a wider world.

The same liberation from the old ways is also seen in the development of Greek political life. By the fifth century B.C.E., Athens had its first Assembly, with full legislative powers, meeting forty times a year and open only to free-born adult Athenian males. No property or class qualification was necessary to qualify for election to the Assembly, and it was every citizen's right to address it.

Athens of the fifth century was run by public-spirited members of the male, nonslave community. From 30,000 free citizens, no fewer than 2,000 offices had to be filled. Only a highly developed system of written record keeping made this possible. Each new official could read the work of his predecessors and learn from them. The new polity relied above all on a literate citizenry that could read public notices, as well as laws and decrees written on stone. One example of this was the practice of ostracism, the process by which a man could be banished from Athens. The decision to do so required at least six thousand citizens to read the charges against him and then write his name on a fragment of pottery (the Greek word for which was "ostraca").

Since all citizens could not be available at one time, at short notice, there must have been well over six thousand literate men in Athens. Most estimates are of at least ten thousand, with a similar number in other Greek centers. This would have represented at least 10 percent of upper-class males. By comparison, in Sumeria and Egypt, it is extremely unlikely that as many as 1 percent had been capable of reading or writing.

The new Greek "democratic" political structure seems to have been one direct outcome of the way alphabetic thought encouraged the discussion of maturing ideas. But it was democratic only in a very limited sense, since there was no universal franchise in Greek society. Above all, the process involved only citizens, and no member of the enslaved majority of the community could vote. So in one sense Greek democracy was little more than a larger version of the coterie with access to the axemaker gifts who had always run society. But it gave the modern world a model on which to build.

Not everybody was happy about how the alphabet might affect thought, and some of the most critical views about this came from Plato. He issued a warning and expressed his concerns about how alphabetic thinking might affect our view of knowledge in an admonition that would go largely unheeded until the late twentieth century:

> It will produce forgetfulness in the souls of those who have learned it, through lack of practice at using their memory, as through reliance on writing they are reminded from outside by alien marks, not from inside, themselves by themselves: you have discovered an elixir not of memory but of reminding. To your students you give an appearance of wisdom, not the reality of it; having heard much, in the absence of teaching, they will appear to know much when for the most part they know nothing, and they will be difficult to get along with, because they have acquired the appearance of wisdom instead of wisdom itself.

It is perhaps too easy to view the rise of Greek thought, epitomized by Aristotle, as the first magnificent attempt to free the human mind from the grip of thousands of years of ignorance and blind ritual. But this view is itself constrained by the fact that what happened twenty-five hundred years ago in Greece shaped the way we ourselves see those events. Our thinking is the product of the Aristotelian system of logic, which itself was designed to prevent the anarchy made so frighteningly possible by the alphabet and made so seductive by the Sophists.

Logic cut at the root of free thinking before it could become anarchic, or develop in whatever alternate form, and would go on doing so over the following two thousand years. It would take that long for another revolution in language technology to offer the human mind a second chance. In the meantime the "cut-up-manipulate-and-sequence" process was so powerfully entrenched in our thinking that within a few centuries of Aristotle and Plato it would survive even what appeared to be the end of the world.

II.

CUTTING UP THE WORLD

Chapter 4

FAITH OF POWER

By the time Rome emerged as an imperial power, the axemakers had provided the means for a small elite to live in relative comfort and order and for the majority to involve themselves in a myriad of different activities. In the fifth century, when Rome fell, to its citizens the end of civilization seemed near.

Once again axemakers came to the rescue. This time their gifts, in the form of Classical knowledge preserved almost intact throughout the Dark Ages, would place in the hands of a single central authority more power over more people, who would conform to more rules of behavior more extensive and constraining than any that had gone before. If the Mesopotamians defined the social structure and the Greeks shaped thought, the new constraints that came in the early Middle Ages would narrow the individual's options even further. These axemakers' gifts would make it possible for leaders to control their followers' most fundamental personal beliefs. This was to develop over the centuries following the end of Rome, whose administration would act as a model for the regime that picked up the pieces after the Empire fell.

With the help of Alexandrine Greek knowledge, the Romans had been the first to run a highly centralized Empire, supported by extensive use of technology and extending over continental distances. At its peak, Rome ran everything from Scotland to the Sudan, and from Portugal to Iran. And the biggest empire the world had ever seen existed to serve only one purpose: to protect and support the central authority. To that end, the Romans used the science and technology of their clever Greek subjects to build and maintain a massive bureaucracy, unparalleled in its influence and extent, perhaps, until the modern Internal Revenue Service.

The Roman propaganda machine welded together a highly diverse set of cultures across Europe, the Near East, and North Africa, so as to strengthen the power of Rome and to support its overall military and political aims. State administrators made use of the visual arts for social control throughout the empire. Caesar and his successors used all the media forms developed in the ancient world to control a vast area of conquest containing thousands of different tribes and languages. Emperors took great pains over their portraiture on coins, showing themselves as war-lords, priests, and divine protectors. Everywhere the Romans had gone, they recreated a Roman environment by building towns of standard ground plan. Regular parades and festivals idealized the state and the emperor, and all the main roads to Rome were lined with "triumphs," the equivalent of giant stone billboards, carrying ads selling the virtues of the Roman way of life.

One of the greatest works in Roman literature, Vergil's *Aeneid,* was written for and sponsored by the minister for propaganda, Maecenas. The Roman poet Horace gave patriotic sentiment a slogan so enduring and powerful that it still appears on dead soldiers' gravestones today: *Dulce et decorum est pro patria mori* ("It is sweet and honorable to die for one's country").

In terms of specialist knowledge, Rome contributed little that was new, though Roman engineers built a network of roads to hold the Empire together that still exists two thousand years later. Possibly due to the need for tight social control in order to manage on such a geographically large scale, the Romans had concentrated on consolidation rather than change.

But by the fifth century it seems probable that the tax burden laid on the imperial provinces to pay for an increasing number of do-nothing bureaucrats and aristos in Italy had weakened local loyalties in the provinces to such an extent that they were a pushover for the first wave of invaders to come out of the Eastern European forests.

And come they did: Goths, Germans, Vandals, Visigoths, and Huns, in an irreversible tide.

After early failure to hold them back, attempts were made to co-opt them into "federated" citizenship. In a few places this trick worked for a while, and there were pockets of late Roman civilization surviving in idyllic areas like Ausonia in Southwestern France as late as the sixth century. But eventually the legions retreated and the stone roads grew over with weeds and bracken. The Saxons and Norsemen and Hungarians knew how to manage highway upkeep, but there simply wasn't anywhere to go and nothing to manage, so the infrastructure sank into disuse and decay, marble was stripped from buildings for lime, and stone was removed to build crude dwellings often in the shadow of the Roman villas the stones had come from.

Throughout the failing Empire, as disease and starvation took its toll following the breakdown of Roman organization, food production became, like everything else, a matter of producing only enough for local use. The broad acres of Roman cultivation gave way to tiny strip fields around villages that were hemmed in by the wall of forest growing back everywhere. Huddled in small clearings were small manors and villages whose inhabitants rarely went farther afield than a half-day's ride or walk so as to be back safe by nightfall.

Within only three or four generations, the Romans had become dimly remembered superbeings, their mark on the landscape commemorated in places named after the now incomprehensible command structure they had wielded: "Street," "Forum," "Camp." Gradually the lights went out all over Europe when there was no more oil for the lamps and no need for illumination at night.

In this world of flickering shadows cast by an occasional rush torch, ghosts and bandits roamed the vast, dark spaces between the tiny settlements. But here and there across the continent were pinpricks of flame, still burning steadily behind thick stone walls, within which Roman life of a much-reduced quality continued. Inhabiting these scattered islands of light in a sea of darkness there were a small number of men, such as Boethius and Cassiodorus, who could still read and write, philosophize and argue, remember what had gone before. They could even communicate with others of their sort, in other places of sanctuary like their own.

This was the next generation of axemakers, holding on desperately to the ancient knowledge and slowly putting together the pieces of a distant, future world that might be once again peaceful and productive, and that would, when the chaos had receded, be ordered according to the vision of the church. These were Christian monks, keepers of what little knowledge was left in the ruins of the West. Beginning with the founding of about forty houses of Celtic monasticism scattered throughout France and Italy, these enclosures of axemaker activity were transition points from which the knowledge of the ancient world would nourish and give birth to mature medieval thought. The very names of these monasteries read like an axemaker roll call: Jumièges, St. Gall, Bobbio, Luxeuil, Ripon, Wearmouth, Jarrow, Bangor, Kells, Corbie, and later Lorsch, Reichenau, and Fulda.

The model for their world-to-come had, ironically, been inspired by false reports of the sack of Rome by Alaric the Hun, back in 411 C.E. Supposedly he had been a barbarian seeking vengeance for his pagan gods. Some Romans also blamed Christians for the end of the empire. People claimed that the ancient gods, offended by the late empire's formal adoption of the new faith, had withdrawn their protection from the Eternal City.

Augustine, Christian bishop of Hippo in North Africa, spent thirteen years on his response to this attack. The result, a work called *De Civitate Dei* ("The City of God"), defined the attitude that would direct Western society through the centuries of confusion that lay ahead.

Augustine expressed the "end-of-the-world" escapist feeling of the time. Beyond the realm of the senses was a spiritual and eternal world of truth that was the goal of all man's strivings. The way to enter this divine world was not by examining the external world with the senses (indeed the social chaos made such investigation impossible), but by turning inward. Truth came not from the external world or from the mind, both of which were impermanent and unreliable, but from the illuminating presence of God.

Augustine divided people into those favored and those outcast, those who dwelt in the "City of God," who would live eternally with the deity, and those in the "Earthly City," who would be condemned to permanent torment with Satan. The Church was the earthly representative of the Kingdom of God and, according to Augustine, would one day wield supreme power in a theocratic society. Augustine's ideology was a tool whose potential the church would use increasingly over the following thousand years to attempt to manipulate and control the secular rulers of Western Europe and through them their subjects.

In the dying years of Rome, the Christian hierarchy modeled their organization on the Imperial administration, where groups of city governments formed provinces and groups of provinces formed vicariates. In the church, the basic unit became the diocese ruled by a bishop. Dioceses were grouped together to form church provinces presided over by archbishops, and provinces were grouped under the direction of metropolitan archbishops, or primates. Ruling the metropolitans were the patriarchs of Rome, Constantinople, Antioch, Alexandria, and Jerusalem.

This tightly knit structure lasted through the centuries of darkness that followed the fall of Rome because its members had the means to keep in contact and share what little knowledge survived the cataclysm. Taking as their text Daniel 12.4: "Many shall run to and fro, and knowledge shall be increased" (or perhaps taking a look around at the crumbling Empire), the early medieval church organized special congregations of priests and lay people to repair local roads, build bridges, set up messenger relays, and even establish hostels for travelers.

The church's magic power was also able to persuade people that if they visited holy shrines they would be granted numinous contact with the relics of Saints. These faithful travelers also acted as messengers for the church, whose lines of communication were kept open at most times throughout this entire period. A widespread and well-organized message network, operating from bishop to bishop, was set up by Pope Gregory the Great in the seventh century.

A hundred years later, St. Boniface used priests to carry his regular and numerous letters from Germany to England and Rome. In them, Boniface referred to items he either wanted delivered or had already received. While he was in Germany, books were sent to him by Abbess Eadburga in England; Boniface asked for details of any books in the library of Abbot Duddo that might be useful; he ordered a copy of the epistles of St. Peter written in gold letters to impress his congregation with honor and reverence; late in life he sought a copy of the Prophets, written out in large letters and without abbreviations, because of his failing eyesight.

The ability to read and write and to communicate over distance raised the Christian hierarchs to an extremely powerful position over illiterate kings and princes, who relied totally on the clergy to help them administer their territories. This was when new phrases came into the language, such as "auditing" accounts and holding "hearings," where oral evidence was presented because most of those involved, including the highest in lay society, were illiterate and could only understand the spoken word. But when a cardinal corrected the Latin of the Emperor Sigismund he replied, *Ego sum rex Romanus et super grammatica* ("I am the Roman Emperor and I am above grammar").

It was easy for the church, primarily through its monastic communities and bishops, to control an illiterate world. By the early Middle Ages, Roman state schooling had vanished and nothing replaced it that might compete with the educational system controlled by the church. Knowledge was now in the hands of a tiny fraction of the population, it was exclusively religious in purpose, and gave the church a monopoly of control over those aspects of social life that required literacy and learning.

Pope Gregory made art a propaganda tool. He said:

> Pictorial representation is made use of in churches for this reason: that such as are ignorant of letters may at least read by looking at the walls what they cannot read in books. To the end that both those who are ignorant of letters might have wherewith to gather a knowledge of the history and that the people might by no means sin by adoration of a pictorial representation. . . . For what writing presents to readers, this a picture presents to the unlearned to behold, since in it even the ignorant see what they ought to follow. In it the illiterate read. Hence, and chiefly to the nations . . . a picture is instead of reading.

In the later Middle Ages, art would be used variously to publicize papal authority over secular rulers. Emperors and anti-popes are repeatedly shown trodden underfoot by a triumphant supreme pontiff sitting on an ever-larger throne. Much later it became common for the popes to be portrayed somewhere in every work of art they commissioned.

As far as scholarship was concerned, for the moment it was time for intellectual consolidation rather than new knowledge, so the church tried to preserve what it could with vast compilations of axemaker knowledge, which were a weird mixture of fact, fiction, and hearsay.

The leading early compiler was a seventh-century Visigothic Spaniard, Archbishop Isidore of Seville, one of whose books, based on Classical Roman and early Christian sources and called the *Etymologies,* was among the most popular texts throughout the Middle Ages. It was an encyclopedia of the world described according to the meaning of the names for things. The subjects included medicine, law, timekeeping and the calendar, theology, anthropology (including monstrous races), geography, cosmology, mineralogy, and agriculture. Isidore's cosmos was geocentric and composed of the four elements: fire, earth, water, and air. He also believed the Earth was a sphere and had a crude understanding of celestial movement.

Isidore was followed by the other great medieval encyclopedist, the English monk Bede (d. 735), who wrote two textbooks on timekeeping and the calendar to help monks work out the hours and days for their prayers and festivals. This would become the basis for *computus,* the principles of timekeeping and calendar control used throughout Christendom.

Neither Bede nor Isidore discovered anything new, but for centuries their collections provided European policymakers with their only source of knowledge about nature, and the church would control access to and use of it. And given the Christian concern with the spiritual rewards of the next world, access was more a system of metaphysical argument about issues like the unknowability of God's name or the Dionysian makeup of the celestial hierarchy. This was a mental environment where the inquiring mind was severely constrained to safe thinking and conservation rather than to rocking any intellectual boats or ground-breaking intellectual discoveries.

Meanwhile the church consolidated its authority by taking draconian action against any and all challenges. When the second-century Gnostic thinkers had proposed a way to salvation through scholarship and self-knowledge, their books were burned and they themselves proscribed. In an echo of cut-and-control strategy from the prehistoric use of gifts such as the Montgaudier baton, the church consolidated power with a tightly controlled canon of official knowledge, sourced in a single text, the Bible, which was accessible only to literate, ecclesiastical hierarchs.

To increase revenue, the church authorities introduced a religious tax, the tithe (tenth), modeled on Old Testament enactments and first authorized by the Carolingian King Pepin the Short (741–68), who ordered that everybody (unless exempted by the pope, as in the case of religious orders) should pay a tenth of their income to the local church where they received the sacrament. This was the first universal tax in European history and greatly aided a papacy in sore need of financial aid.

The power of the church lay, above all, where all social power had always been, in a centralized command structure. The pope in Rome, at least in theory, was the single central authority to whom total obedience was due. The royal power of kings and princes was recognized, but church teaching imposed restrictions on its extent.

As early as the fifth century, Pope Gelasius I (492–6) had said: "There are two [principles] by which this world is mainly ruled, the sacred authority of the popes and the royal power. Of these two, the weight of the priests is much more important, because it has to render account for the kings themselves at the divine tribunal." Although this was basically an admission that society indeed needed both the monarchy and the papacy, it contained the future seeds of more radical papal thought.

The pope's leadership was, however, often contested. Gregory VII achieved a decisive step to strengthen the authority of the church with his *Dictatus Papae* of 1075, where he declared among other things: "Only the Roman pontiff is justly called universal . . . He is the only one whose name should be pronounced in all churches." Later, in the twelfth century, the pope's title changed meaningfully from "vicar of St Peter" to "vicar of Christ." Certain powers eventually belonged only to the pope: he was the final court of appeal in ecclesiastical cases, his permission was needed to set up a new province in the church, and he alone could remove an archbishop from office or exempt clergy from the jurisdiction of their bishop, a direct threat to the power of royal episcopal appointment. The pope could also make written rulings binding on the whole church, appoint legates to represent him in any part of the church, and summon a General Council.

Growing papal power was slowly becoming evident in the West. From the seventh century, it became a requirement that every newly elected archbishop should visit Rome to be invested with a pallium, a woolen scarf he was obliged to wear, when officiating, as a sign that he was in communion with the Roman See.

Parallel with this, the papacy began to take over the old, imperial function in the West, particularly from the eighth century onwards, thanks to the counterfeit Donation of Constantine. According to the document (discovered to be a forgery only in the fifteenth century), in the fifth century, the Emperor Constantine had given the city of Rome (and, by implication, control of the whole of the West) to the pope, when he had transferred the imperial headquarters eastwards to Constantinople. The "donation" authorized the pope to wear the diadem and papal insignia and conferred senatorial rank on the Roman clergy. From the time of Paschal II in 1099, the popes were crowned on accession, and after Gregory VII their "enthronement" at the Lateran was accompanied by the donning of the imperial red mantle, possession of which, in the case of rivalry between popes, conferred legitimacy. On the other hand, imperialists could use the Donation to "prove" that the papacy exercised its control *through* their favor.

Not only did the papacy try to remove the church from subservience to lay authority, but it also proclaimed itself to be superior to the secular hierarchy. In response, European kings and emperors began to invest themselves with a religious character, one at least sacred if not priestly. They started by having themselves anointed at coronations, which took the form of religious ceremonies that declared them to hold power as the "Lord's anointed." The actual anointing was, of course, done as often as possible by the pope. In this way, Rome attempted to underpin royal power by endowing kings and emperors with a sacred character and making it obligatory for all subjects to submit faithfully and with blind obedience to them because (in an echo of Mesopotamia) "he who resists this power resists the order willed by God." But faithful submission was a two-edged sword and could be turned against the papacy by the royal power.

The church council called to Paris in 829 even defined the duties of kings in terms that were taken up and developed two years later by Jonas, bishop of Orléans, in his *De institutione regis,* which remained the model for royal instruction throughout the Middle Ages. At the council the bishops announced:

"The royal ministry consists especially in governing and ruling the people of God in fairness and justice and in seeing to the provision of peace and concord. Above all, the king must be the defender of the churches, of the servants of the God, of widows, orphans, and all other poor and needy people."

Ecclesiastical control, based on their literate abilities, inserted the clergy into every aspect of secular life. Bishops and abbots received grants of land from kings and noblemen, although royal appointment often placed them in an inferior position to the monarch. But it did enhance their political and economic power in all the kingdoms of the West, giving them power-of-landlord over thousands of peasants. Throughout this period, bishops and abbots sat in royal councils, were influential in drafting secular law codes, and took a major part in affairs of state. During the ninth and tenth centuries, churchmen also frequently became involved in military organization, since from the ninth century onwards a number of conditions began to be attached to grants of land, often including an obligation that the church recipients should muster a specified number of troops when requested for their benefactor's service.

By the eleventh century the church's grip on Western society was firm, although not uncontested. Churches had been established in all major areas of settlement throughout northern Europe, and this made possible the growth of a parish system. Everybody living in a town or village in western Europe had a local church.

The church then took social control to new levels, achieving unprecedented mastery over the thoughts and feelings of every individual through one of the most effective systems for social discipline ever devised: the confession. By the twelfth century, any sins or offenses committed against church doctrine had to be privately divulged to a priest, and failure to do so could lead to punishment, even to the ultimate sanction of excommunication from the Christian community, which would deprive the guilty party of all forms of protection under civil or canon law. The practice probably began in Celtic monastic penitential practices, where a monk or hermit confessed his sins to his "soul-friend," as the texts called this moral guide.

Slowly the system became commonplace to the point where the Lateran Council of 1215 decreed that everybody had to confess once a year to the parish priest. This was one of the most important steps taken in over a thousand years to enforce the Christianization of hearts and minds. Mental control, first exercised over monks by other monks, came to be exercised over everybody by the secular clergy responsible for spiritual care. From this point on, nothing would remain hidden from the church. However, as things turned out, the church would need every control system it could marshal to deal with the threat (to Christendom and above all to papal mind-control) that had been building up in the Middle East since the seventh century.

By this time, Alexandrine Greek knowledge had begun to transfer to Islam, where it would be processed before returning once again to Western culture. The transfer was triggered by a heretic Christian sect, the Nestorians, who had been expelled from Byzantium centuries earlier and had wandered Asia Minor until settling at a place in Jundishapur in the mountains of southern Iran, a few miles from what would become the site of the first Arab capital, Baghdad.

In seventh century Baghdad, Caliph Al Mansur was looking for a cure to his gastric problems and sent servants to the Nestorian monastery to ask for medicines. His retainers reported the presence in the monastery of an immense library. Al Mansur subsequently discovered that the Nestorians had preserved virtually intact the work of all the major thinkers at the Alexandrine Museion as well as that of their Classical Greek predecessors. The caliph and his successors ordered almost all the texts translated and found themselves in possession of a treasure trove of Greek axemaker knowledge.

The transfer of the Greek data reached its height during the eighth and ninth centuries in Baghdad under the Abbasid Caliphs, with the translation of Aristotle and Plato, Hippocrates and Galen, Ptolemy, Euclid and Archimedes, Apollonius, Aristarchus, and others. Nothing was absorbed into Islamic culture until it had been checked for theological acceptability. This work was done in libraries, hospitals, and observatories, and it stimulated Arab axemakers to their own investigations of the world. Astronomy told them the hours of prayer and the direction of Mecca; medicine was a valuable applied science, linked to astronomy through the astrological nature of treatment; philology aided the analysis of sacred writings.

Gradually, however, Islam drew a distinction between religious subjects (law and religious custom), the subjects to be used in the service of religion, such as astronomy and grammar, and the secular sciences of math, astronomy, and medicine. Islamic societies turned Greek theory into applied technology that would help them and their clerical masters to survive and prosper. They achieved important advances in hydraulics and applied them to irrigation systems so that their deserts bloomed with the magnificent gardens of their rulers' palaces.

The highly centralized nature of Islamic society, which placed tight constraints on the individual's freedom of intellectual movement, made innovative thinking possible but its application strictly controlled. The same was true of medieval Chinese society, where at this time other axemaker knowledge was being generated, which would, like Islamic advances, eventually find its way West. In China, the state controlled all activity, and the comprehensive social organization originally required for irrigation and other large-scale public works gave Chinese life a collective character.

All individual activity in China was subordinate to the common good and thus defined by the bureaucrats. From earliest times, power had rested exclusively in the hands of a shaman ruler, who was the divine son of heaven. He was supported by an extensive, all-powerful Mandarin bureaucracy, entry to which was by merit and most of whom followed the teachings of the fifth-century B.C.E. thinker Confucius. To the Mandarins, Confucian thought was the "Great Way of Life," whose tenets controlled all social and political activity, as well as keeping tight constraints on the activities of the free-thinking analytical mind.

The Confucian view was another good example of the self-fulfilling nature of axemaker processes. According to the Confucian tenet, the only purpose of education was to prepare for service to the state, so the aim of the educated man was to involve himself primarily in the maintenance of stable government. No knowledge was derived from supernatural revelation; rather it was reached through the use of reason, which also made explicit the guidelines for ethical conduct, which in turn was defined by the state.

In this closed loop, there was no way for scientific theory to become technological practice because the state decreed that no contact was permitted between one discipline and another, so theory was not expected to relate to practice. The Mandarins believed the most powerful tool for social management was classification and record keeping of everything and everybody, so everything was categorized, and the application of knowledge was only permitted within its own category. While all the requisite information was available, a revolution did not happen in China because it was all separated.

When the immense repository of Islamic (and through Islam, Chinese) and Alexandrine Greek knowledge finally reached the West at the time when Europe first made contact with the Arabs in Spain, Sicily, and Jerusalem, it would first of all put unprecedented power of cut-and-control into the hands of the Catholic leadership, thanks to the Christian belief that they had a God-given right to subjugate the world.

According to both the Old and New Testaments, man had been given dominion over nature. Genesis said: "Every living thing shall be meat for you . . . the fear of you and dread of you shall be upon every beast of the Earth . . . Into your hand they are delivered . . . Have dominion over the Earth and subdue it." When these early statements were originally made, they were probably intended to regulate and maybe celebrate what had been happening since the axemakers had made possible settlement in the prehistoric Levant and principally to memorialize the domestication of animals and the first agriculture.

In many other religions, nature was divine, or it shared divinity, but Christian doctrine gave humankind a position separate in nature from the rest of created things. Greek cosmologists had also shared this view that nature was not sacred, so when Arab translations of Aristotle's work arrived in the Christian West in the early Middle Ages, his statement that animal life only existed for man's sake added extra authority to Christian practice.

The dominant Christian view was that since animals and plants did not have souls, this precluded their eligibility for humane treatment. Manipulation of nature (which could include the enhancement of its value and beauty) was mankind's right and duty because "improvement" of the world involved the exercise of power derived from God for that very purpose.

The medieval Christian believed in Aristotle's "Great Chain of Being," the hierarchical structure created at Creation by God or, as Aristotle would have put it, the "Prime Mover." The Great Chain linked all species, one to the next, from the simplest organism all the way up to humans and angels and was based on the concept that lower forms existed only for the sake of higher forms.

Eleventh-century Benedictine monks were among the first systematically to apply these views of nature to their daily life and to begin a process of "improvement" of nature, which would be reflected in axemaker activity for centuries to come. The Rule of St. Benedict had ordered members of the Order to seek out monastic sites "far from the haunts of men," in wild and isolated places, and then to apply their knowledge to cultivation of the land so that it would provide enough food to support them.

One particular family of Benedictines, the Cistercians, whose motto was "Work is prayer," succeeded best of all in this task. Most of what little technology survived the centuries after the fall of Rome emerged in the Middle Ages from Cistercian monasteries that were more like small factories, filled with water-driven looms, mills, saws, grinding stones, and trip-hammers.

In the twelfth century, St. Bernard of Clairvaux in France described the landscape at his Cistercian monastery as having been "given meaning," since human ingenuity had brought order to the wilderness and had dammed the river, diverting its flow to drive the monastery's waterwheels:

> The [river] Aube . . . passes and repasses the many workshops of the abbey, and everywhere leaves a blessing behind it for its faithful service.
>
> The river climbs to this height by works laboriously constructed, and passes nowhere without rendering some service, or leaving some of its water behind. It divides the valley into two by a sinuous bed, which the labor of the brethren, and not Nature, has made, and goes on to throw half of its waters into the abbey, as if to salute the brethren and seems to excuse itself for not coming in its whole force, the canal that receives it being too small for it. If sometimes the stream, swollen by an inundation, rushes on with violent current, it is stopped by a wall, under which it is obliged to pass and so turned back upon itself, meets and checks the descending stream. As much, however, as the wall, like a faithful porter, allows to enter passes on at once to drive the wheels of the mill; there, lashed into foam by their motion, it grinds the meal under the weight of the millstone and separates the fine from the coarse by a sieve of fine tissue.
>
> A little farther on, in the next building, it fills a boiler and is heated for brewing, that drink may be prepared for the brethren, if it should happen that the vintage should not respond kindly to the labor of the vine-dresser; so that, in default of the juice of the vine, the want may be supplied by the extract of grain. But not even yet is its usefulness completed, for the fullers call it to their aid who labor beside the mill; sound reason requiring that, as in the mill, care is taken for the food of the brethren, so by these their clothing should be prepared. But the river does not hesitate nor refuse any who require its aid; and you may see it causing to rise and fall alternately the heavy pestles, that is to say, hammers, or wooden, foot-shaped blocks (for that name seems to agree better with the treading-work, as it were, of the fullers) and so relieves them of the heaviest part of their labor . . . How many horses would this labor tire! Of how many men would it weary the arms!

Twelfth-century Cistercian monasteries were the most advanced technical complexes on the European continent, with the most developed agricultural techniques and the most productive factories and mines. It was their dynamic "go forth and improve" doctrine that was eventually to give secular authorities of the late medieval West the technology to achieve efficient means of social control.

One of the new control systems arose out of the liturgical needs of the monasteries in Northern Europe, where monks needed to know what time it was because there were specific rules scheduling daily collective prayer for the souls of the multitude that it was a monk's duty to perform. There were seven set times at which prayers were to be said, some of them in the middle of the night. At first, water clocks and candles had served to indicate when the monastery bells should be tolled to mark prayer time, but water clocks froze in winter and candles blew out.

For the monastery overseers in the monastic proto-factories, timekeeping was also an indispensable method of organization. So it may have been the expansion of this technology-oriented religious order of axemakers that intensified the search for a better form of timekeeping and spurred the development of the mechanical, weight-driven clock in the thirteenth century.

The gift of the clock immediately made possible new forms of wider, more effective marshaling of social forces. Demand for clocks from royal courts and from the growing number of towns throughout Europe was overwhelming. Town clocks gave guilds and governments the means to regulate all behavior. In Brussels, textile workers rose at a dawn bell, weavers and twisters ended their day with an evening bell, and there was a special clock for cobblers. In 1355, in Amiens, France, the city government would issue an ordnance "concerning the time when the workers . . . should go each morning to work, when they should eat and when to return to work after eating; and also in the evening when they should quit work for the day," and they used a special bell for this purpose.

At the same time as this control-technology was spreading out of the monasteries into the urban community, the church was also defining the permitted uses of the new investigative techniques spreading out of the scriptoria, where Arab and Greek manuscripts were being translated and copied.

The earliest translations from the Arabic (several treatises on mathematics and the astrolabe) were made late in tenth-century Spain. A century later, Constantine, a Benedictine monk from North Africa, made his way to the monastery at Monte Cassino in Southern Italy, where he began to translate medical treatises from Arabic to Latin, including the works of Galen and Hippocrates. These would lay the foundations of medical literature on which the West would build for several centuries.

By the first half of the twelfth century, translation had become a major specialist activity, with Spain as the geographical focus since it had had centuries of brilliant Arabic culture, a plentiful supply of Arabic books, and communities of Christians (Mozarabs), who had been allowed to practice their religion under Moslem rule and who could now help to mediate between the two cultures.

As a result of the Christian reconquest of Spain, centers of Spanish Arabic culture and libraries fell into Christian hands. Toledo, the most important, fell in 1085. Greatest of the translators from Arabic to Latin was Gerard of Cremona, from Northern Italy, who went to Spain in the late 1130s or early 1140s in search of Ptolemy's *Almagest,* a work he had been unable to locate elsewhere.

Gerard found a copy in Toledo, stayed in the city to learn Arabic, and eventually translated the work into Latin. However, he also discovered texts on all kinds of subjects, and over the next thirty-five years produced translations of at least a dozen astronomical texts, seventeen works on mathematics and optics, fourteen works on natural philosophy (including Aristotle's *Physics, On the Heavens, Meteorology,* and *On Generation and Corruption),* and twenty-four medical works.

Translation from the Greek, which had never entirely ceased, thanks to the Byzantine occupation of parts of Italy, now dramatically accelerated, especially in southern Italy and Sicily, where there had always been Greek-speaking communities and libraries containing Greek books. A series of important works on mathematics and mathematical science appeared in Greco-Latin translations around the mid-twelfth century: Ptolemy's *Almagest* and Euclid's *Elements, Optics,* and *Catoptrics.* Greco-Latin translations continued in the thirteenth century, notably in the work of William of Moerbeke, who set out to provide Latin Christendom with a complete and reliable version of Aristotle's works, revising existing translations from the Greek where that was required, as well as translating some mathematical works by Archimedes.

These translations made practical knowledge more available to Western ecclesiastical authorities. Medicine and astronomy came first, in the tenth and eleventh centuries; early in the twelfth century the emphasis seems to have shifted to astrological works, along with mathematical treatises needed for the successful practice of astronomy and astrology. Medicine and astrology each rested on philosophical foundations, and it was at least partly to recover these that, from 1150 on, attention shifted to the physical and metaphysical works of Aristotle. But once the full scope of Aristotle's work became known, it became clear that his philosophical system was applicable to an enormous range of issues to be dealt with in the schools and in the new universities.

To Western policymakers, mastery of the newly rediscovered Aristotelian logic gave the almost magical ability to increase knowledge without end and, above all, to provide a damage-control system with which to limit the destabilizing effects of the new Arabic data. As in Greece, the gift of reason might initially offer heady intellectual prospects, but in the end it would act as a effective brake on freedom of thought.

The excitement over Arabic knowledge is clear from the first reports of contact with it. In the early twelfth century, the Englishman Adelard of Bath returned from Arab Sicily to exhort his fellow monks to: ". . . think for yourself. For I have learned something different from my Arab masters, with reason as a guide. You however, taken captive by authority, are led by a halter."

Adelard wrote two books that made a big impact on his fellow-Europeans. In them, he said that all authority should be subject to reasoned questioning. Perhaps his most powerful statement was: "The visible universe is subject to quantification and is so by necessity. . . . If you wish to hear more from me, give and take reason, because I am not the kind of man to satisfy his hunger on the picture of a steak!"

Peter Abelard joined the new thinkers around 1110. He rapidly became a central figure, and in the first decades of the century he was the most conspicuous scholar in Europe. His prime interest lay in marshaling the arguments of authorities in support of and in opposition to disputed points, a technique already in limited use chiefly among legal scholars and theologians and known as the method of *pro et contra.*

Abelard made possible greater precision and more skillful application in the dialectical treatment of disputed theological topics with his influential *Sic et Non* ("Yes and No"), where he set out 157 questions and answers on faith and morals. The work exposed internal inconsistencies in much of the theology of the day and produced a new awareness of the need to apply the principles of logic to human experience and to distinguish logical from metaphysical discourse.

These kinds of inquiries were rapidly becoming a major feature of the new learning of the cathedral schools, and gifted scholars like Abelard proved strong attractions to the increasing number of students in the growing cities. Of course, the dialecticians of these schools were not seeking to destroy faith or to overthrow the established order based on ecclesiastical supremacy. On the contrary, they were convinced that their efforts would strengthen the basis by which the church's absolute values of life could be secured. Abelard's famous dictum: "Through doubt we are led to inquiry, through inquiry we reach the truth" was an attempt to underpin both faith and the established Christian social order.

The new learning became the central feature of intellectual life in the thirteenth century, setting an agenda that would preoccupy the best scholars of the century. The task of these Christian axemakers was to master the new knowledge, organize it, assess its significance, discover its ramifications, work out its internal contradictions, and make it available (wherever possible) for approved application to existing intellectual concerns. Above all, the requirement was make it socially safe. The new translations from Arabic were enormously attractive because of their breadth, their intellectual power, and their utility, but their pagan origin meant they contained material that was theologically dubious.

Most of the translated texts were, however, considered harmless enough, and the very fact that a manuscript had been translated at all meant that its usefulness outweighed its potential for social disruption. Technical treatises on all manner of subjects (mathematics, astronomy, statistics, optics, meteorology, and medicine) were received with unqualified enthusiasm because they were clearly superior to anything previously available and contained no unpleasant philosophical or theological surprises. In this way Euclid's *Elements,* Ptolemy's *Almagest,* al-Khwarizmi's *Algebra,* Ibn al-Haytham's *Optics,* and Avicenna's *Canon of Medicine* were peacefully added to the intellectual armory of Europe's ecclesiastical and (to a limited extent) secular elite.

But problems arose in broader subject areas that threatened to clash with theology, like cosmology, physics, metaphysics, epistemology, and psychology. Central to these subjects were the works of Aristotle and his commentators, who successfully addressed a multitude of critical philosophical problems, while promising untold future benefits from the general use of their methodology.

One of the most eminent of Christian scholars to be influenced by the new material was Albertus Magnus, a thirteenth-century Dominican friar, who taught at the University of Paris. Inspired by Aristotle's ideas, he traveled through Europe, asking questions of the kind of craftsmen (fishermen, hunters, beekeepers, and bird catchers) that the church usually ignored and looking at the world with new eyes. He produced two books on botany and zoology filled with descriptions of nature, which have a freshness and an immediacy quite unlike anything written in the previous thousand years. His remark about the bird that rose from fire ("The phoenix owes more to mystical theology than to nature") was an extraordinarily modern statement for the time.

At a moment when ancient texts were the source of all authority, Albertus put forward a revolutionary new methodological principle: "There can be no philosophy about concrete things. . . . In such matters, only experience can provide certainty." Albertus responded to the Aristotelian rationalism and commitment to the application of philosophical method in all areas of human enterprise by proposing to distinguish between philosophy and theology on methodological grounds and to find out what philosophy alone, without any help from theology, might demonstrate about reality. This was to prove to be a breakthrough.

A French contemporary, William of Conches, took things further and called for the objective investigation of faith and philosophy. The French scholar Thierry who lived in Chartres (the center for the new thinkers, who were called by their contemporaries "the modern ones") analyzed Genesis from the point of view of the natural processes described in the text and questioned the extent to which the descriptions should be taken literally.

The key issue, however (and one that would threaten the very foundations of the church), was that if logic and reason were to be applied to faith, what would become of miracles such as the Virgin Birth? What would become of faith? Scholars had to be careful to maintain the delicate balance between belief and unbelief called for by this kind of thinking.

William of Conches protested that his philosophical position did not detract from divine power and majesty: "I take nothing away from God; all things that are in the world were made by God, except evil; but he made other things through the operation of nature, which is the instrument of divine operation." Study of the physical world enabled men to appreciate "divine power, wisdom and goodness," so searching for secondary causes (natural processes) was not a denial but an affirmation of the existence and majesty of the first cause (God).

The reality of miracles could in this way be reconciled with nature by acknowledging that miracles represented genuine suspension of the usual laws, taking for granted that these suspensions had been planned by God from the time of Creation and were built into the cosmic machinery. In this way, miracles could remain perfectly natural, in the larger sense. It was possible to talk about a fixed natural order without infringing on divine omnipotence and freedom, by arguing that God had unlimited freedom to create any world he chose, but that he had, in fact, chosen to make the world as we find it. And now that it was completed, God was not going to alter it. This approach was to become crucial to Christian authority.

In the second half of the thirteenth century the controversy over the new views began to involve many of the recently established universities of northern Europe, at the time dominated by, or owing some degree of allegiance to, the papacy. The center of the conflict was Paris, although the row spread to other French universities in Toulouse, Montpellier, and Orleans. The problem was whether or not the universities had the right to study the new texts and especially Aristotle's treatises on metaphysics and natural philosophy, particularly as presented to the West through the writings of the Spanish Islamic commentator of Aristotle's work, Averroes.

The issue was a hot one, because if Aristotelian natural philosophy were to gain acceptance, then the entire metaphysical basis of the church's traditional Augustinian teachings and its claims to religious authority might be challenged, and the way would be open for the development of a completely naturalistic, rational explication of the universe, with obvious danger for the Church. Ironically, in the early thirteenth century, Rome reacted to the new heterodoxy much as Greece had reacted to the Sophists fifteen hundred years before, with a total ban on the teaching of Aristotle.

In Paris, allegations were made that pantheism (roughly speaking, making God a part of the universe) was being taught by masters of arts under Aristotelian inspiration. This resulted in a decree, issued by a council of bishops meeting in Paris in 1210, forbidding instruction on Aristotle's natural philosophy within the faculty of arts. The decree was renewed in 1215 by the papal legate Robert de Courcon. Pope Gregory IX became directly involved in 1231, renewing the ban of 1210 and specifying that Aristotle's books on natural philosophy were not to be read in the Paris faculty of arts until they had been "examined and purged of all suspected error."

In a letter appointing a commission to act on the matter, Gregory wrote: "Since the other sciences should serve the wisdom of Holy Scripture, they are to be appropriated by the faithful insofar as they are known to conform to the good pleasure of the Giver." However, Gregory was aware that "the books on natural philosophy that were prohibited in a provincial council at Paris . . . contain both useful and useless matter" and therefore "in order that the useful not be contaminated by the useless" he ordered the commission to "eliminate all that is erroneous or that might cause scandal or give offense to readers, so that when the dubious matter has been removed, the remainder may be studied without delay and without offense."

Then in 1277 the discussion of anything remotely related to rationalism was forbidden, while Rome looked for a means out of the apparent impasse. The way was found in the person of a Dominican intellectual who had studied under Albertus Magnus in Paris and whose name was Thomas Aquinas.

Aquinas papered over the cracks between faith and reason in his *Summa Theologica.* In it he argued that philosophy examined the supernatural order in the light of reason and theology examined it in the light of revelation. Although reason was used in theology, revelation did not fall within the province of philosophy, and philosophy could not contradict theology because truth could not contradict truth. Human reason could demonstrate some truths of revelation and it could show that other truths were supra- rather than antirational, but faith was a realm in which reason could not hold sway.

For Aquinas, faith and knowledge were, therefore, not mutually exclusive. He said that belief took over at the point where knowledge ended. The goal of both reason and theology was "Being," and although reason could not finally grasp "Being," it could make faith plausible. In this way, he showed that faith and knowledge were not antithetical. Aquinas summed up his view: "To believe is to think with assent."

Aquinas showed the lack of tolerance to opponents that might be expected from a defender of the establishment, justifying excommunication and execution and arguing that since their sin affected the soul, they should be more quickly and severely punished than forgers and robbers. However, the church should admonish them twice, hoping for their return, before excommunicating them and turning them over to secular powers for execution.

With this *Summa,* Aquinas released the full power of the gift of rationalism into secular hands. He bowed to the power of geometry, admitting that God could not make the sum of the internal angles of a triangle add up to more than two right angles. In the future there would be two kinds of knowledge: that which related to revelation (which would be the province of theology) and that which dealt with the natural world (which reason and philosophy could handle).

With this decision the church created another opportunity for axemakers to go forth and multiply. The result would one day become known as "science." But the unshackling of rationalism in this way was only a matter of appearances. No "science" would be free of ecclesiastical control for centuries. Indeed, for most of the time up to the modern world, most scientists would be churchmen, and as late as Darwin science would continue to work in support of established religion.

One of the earliest expressions of the new, more secular view came at the end of the thirteenth century from an English cleric called Roger Bacon, in his *Opus Maius.* Writing about Peter of Maricourt, another traveler to Arab lands who was already famous for his work on magnetism, Bacon said: "What others strive to see dimly and blindly, like bats in twilight, he gazes at in the full light of day because he is a master of experiments. Through experiment he gains knowledge of natural things medical, chemical and indeed of everything in the heavens or Earth."

Bacon's major "scientific" writings were not pieces of natural philosophy but passionate attempts to warn the church hierarchy (in works addressed to the pope) against suppressing the new learning expressed in Aristotelian philosophy and in all the new literature relating to natural philosophy, mathematical science, and medicine. Bacon argued that the new philosophy was a divine gift, capable of proving articles of faith and persuading the unconverted; that scientific knowledge contributed vitally to the interpretation of Scripture; that astronomy was essential for establishing the religious calendar; that astrology enabled man to predict the future; that "experimental science" taught how to prolong life; and that optics enabled the creation of devices that would terrorize unbelievers and lead to their conversion.

There was "one perfect wisdom," Bacon argued in his *Opus Maius,* "and this is contained in holy Scripture, in which all truth is rooted. I say, therefore, that one discipline is mistress of the others, namely, theology, for which the others are integral necessities and that cannot achieve its ends without them. And it lays claim to their virtues and subordinates them to its nod and command." So theology did not oppress these sciences but put them to work, directing them to their proper end.

Bacon's experimental technique, which would give axemakers a new technique for manufacturing knowledge, became known as "resolution and composition." It was a direct descendant of the mode of thought made possible by the alphabet because it applied the cut-and-control analytical method to the solution of problems. "Resolution" defined a complex phenomenon and its causal conditions by breaking it down into the elements or principles involved in its appearance. "Composition" then used this data to show how these causes brought the phenomenon about, thus revealing the conditions that were necessary and sufficient to produce the phenomenon.

The first experiments along these lines were carried out by Bacon and his English contemporary Robert Grosseteste (the first Chancellor of Oxford), as well as Theodoric of Freiburg and others. The aim was to find "mechanisms to make the phenomenon," by experimentally creating the conditions for a phenomenon to exist. Theodoric sprayed water droplets in order to simulate the conditions for a rainbow, then investigated the optical properties of the droplets by creating models of them with spherical flasks full of water and arrived at an explanation of the geometry of light refraction.

Starting in the thirteenth century, the new experimenters began for the first time to refer to nature as if it were a machine that functioned according to discoverable, measurable "mechanisms." In Paris, Nicholas Oresme compared the universe to a clock. The investigators began to describe phenomena as "primary" (physical activity that produced light, heat, or sound) and "secondary" (sensations produced when these phenomena affected the senses).

They were laying the groundwork for an entirely new body of knowledge that would enormously expand the power and influence of institutions and individuals with access to it. In the fourteenth century these new techniques for manufacturing knowledge were still limited to tiny, isolated groups of clerics. But their isolation would end with explosive results a hundred years later, when in 1439 a German goldsmith would get the date wrong. The consequences of his mistake would shake the authority of Rome to its foundations and create an entirely new kind of axemaker.

Chapter 5

FIT TO PRINT

From the time of the first axe, knowledge had conferred power on those who were in a position to be able to make use of it. With each of the axemaker's gifts, from the first "sequentializing" mental effects of language and toolmaking to predictive shaman batons and the bureaucratic potential of Mesopotamian cuneiform script, as well as the analytical force unleashed by the alphabetic world-cutting edge of logic and the mind constraint made possible by the confessional, those institutions and individuals in power were armed with ever-more effective knowledge that they could use to cut and control the natural world and human society.

The next gift would radically change how knowledge was recorded and disseminated. It would also change the nature of knowledge itself, how it could be used, and how many people could have access to it. And in the way that all advances in communication make things more complicated, this gift would break up the monolithic social structure of Christendom and diffuse control outward to many peripheral centers of power. This was possible because, at a stroke, the new gift also increased the number of changemakers.

In 1439, in the German town of Mainz, a goldsmith named Johann Gutenberg found out he had been misinformed about the date of a pilgrim's fair in nearby Aachen. Many of the Mainz townsfolk were supposed to be going to the fair, and Gutenberg had agreed with a couple of investors to make small mirrors to sell to these pilgrims. When it turned out that the fair was to be held a year later than Gutenberg had thought, he revealed to his coinvestors an alternative investment opportunity he had been thinking about for some time: to make individual letters of metal so as to combine and recombine them to print words on paper.

This revolutionary technique had been earlier developed in fourteenth-century Korea, but its use was permitted only to make replacements for court religious texts destroyed by fire. Once the work of replacement was complete, the machinery was also destroyed. Whether or not Gutenberg got his idea from someone who had traveled and heard of the Korean event is unimportant. The fact is that moveable typeface would radically change the world of documentation in the West, replacing, as it did, handwritten manuscripts. A limited amount of printing with engraved wooden blocks had taken place, but its use had been limited almost entirely to illustrations and playing cards. The major limitation of the carved wooden block was that it could only be used to print one image and wore out with use, whereas the secret of Gutenberg's metal typefaces was that they lasted well, reproduced single letters, and were interchangeable.

The effect of Gutenberg's letters would be to change the map of Europe, considerably reduce the power of the Catholic church, and alter the very nature of the knowledge on which political and religious control was based.

The printing press would also help to stimulate nascent forms of capitalism and provide the economic underpinning for a new kind of community. And with the information made widely available by the press, the rapidly growing commercial sector of society would drive all Europe from being a monolithic, backward-looking, medieval culture to a dynamic and complex world power. First of all, however, it would cut up Christianity.

Printing spread across the continent at extraordinary speed. In 1455, there were no printed texts in Europe, but by 1500 there were twenty million books in 35,000 editions, one book for every five members of the population. In 1455, the only printing press in Europe had been Gutenberg's, but by 1500 there were presses in two hundred and forty-five cities, from Stockholm to Palermo.

The printers set up their presses in every university town and major commercial center, and from 1500 to 1600 produced between 150 and 200 million texts. In a special sense, the book was the first modern-style, mass-produced industrial commodity. No innovation in history had ever spread so far, so fast.

Most inventions that create a new world do so by first succeeding in the old one, and so it was in the case of the press, when the printers first put their new, high-tech equipment at the service of the most powerful authority of their time, the Catholic church.

Rome realized that printing could strengthen its social authority through the production and dissemination of thousands of copies of identical devotional books, which would make possible liturgical conformity and obedience on an unprecedented scale. So between 1455 and 1500 over two hundred editions of the Bible were commissioned, together with printed editions of Donatus' Latin grammar, the mainstay of church education since the fifth century. The major devotional text, the *Imitation of Christ,* rapidly became the most popular book in history, second only to the Bible.

Then in 1466 Rome made a move to entrench its power among the growing numbers of literate non-Latin-speakers (in the rising artisan class) by sanctioning the printing in Strasbourg of the first vernacular Bible, in German. The idea caught on, and by 1471 an Italian Bible was on sale in Venice, in 1477 the Delft press had printed a Dutch Bible, and by 1500 there were thirty vernacular editions in six languages. This decision was a major mistake.

By the time Rome realized what effect these vernacular Bibles would have and the way they would reduce universal Catholic power, it would be too late. To start with, the Bibles would have unexpected political effect. They gave permanence to the languages in which they were printed, and in doing so strengthened the unity (and the power of the rulers) of each language community. Between 1478 and 1571, in spite of the fact that Latvia, Estonia, Lithuania, Wales, Ireland, the Basque country, Catalonia, and Finland were all within the economic sphere of influence of another, more powerful language group, these countries retained and strengthened their national identities because they had their own version of the Bible.

Languages in which Bibles were not printed either disappeared or became provincial dialects, subordinate to the politically or economically dominant language group of the area. Without a local Bible to sustain them, the language and political identity of Sicily were subsumed within those of Italy, of Provence and Brittany within France, Frisia within the Netherlands, Rhaetia within Austria, Cornwall within England, and Prussia within Germany. The technology and economics of print production and distribution inevitably also tended to concentrate output on fewer, larger markets, so the printers themselves contributed to a rapid homogenization of the many dialects of Europe into a few major languages.

The political result of these new print-languages, imposed by kings through their control of the presses, would lead directly to the emergence of a new kind of patriot-axemaker. Thanks to printing, the Christian who had belonged to the *oikoumene* of Christendom now saw himself as a member of a group that before print had not existed to any great extent: a nation.

The development of national languages, the loss of a Latin *lingua franca,* and the break-up of Christendom concentrated local control in the hands of independent national leaders. In a speech in the late sixteenth century, Henry IV of France addressed the key issue: "As you speak the French language by nature, it is reasonable that you should be the subjects of a King of France. I entirely agree that the Spanish language belongs to the Spaniard and the German to the German. But the whole region of the French language must be mine."

Monarchs and their governments now began to enforce the local tongue with laws, taxes, armies, and the state bureaucracies that went with them all. And once a state's boundaries had been consolidated in these bureaucratic ways, it gradually became more convenient (for political, economic, and social reasons) to use a single language.

No European used the press more effectively to promote and manipulate this new print-generated sense of national identity than the German Protestant reformer Martin Luther. The presses took his fight with the pope to the streets with astonishing speed. A printed version of his criticisms of the Roman church appeared everywhere in Germany within two weeks of their publication and then all over Europe within a month. Significantly, in 1520 his appeal for support was in a tract entitled: "To the Christian Nobility of the German Nation." It sold 4,000 copies in three weeks, and before the end of the year thirteen editions had been printed.

Later, when he used the presses to produce his Bible, which eventually ran to four hundred and thirty editions, Luther clearly expressed his desire to weld his countrymen into a single linguistic unit that would be easier to influence and control. He said: "I want to be understood as well in Southern as Northern Germany" and in order to do so he standardized German vocabulary and spelling and expunged dialects. The first grammar for the new pan-German language appeared in 1525.

Given the publishers' desire for profitable return on investment, within each new European language efforts were also directed toward the standardization of grammar and vocabulary so as to create a linguistically homogeneous market. Thanks to such efforts by the English printer Caxton, the London dialect became the national language, and in Italy Dante's Tuscan became the "official" Italian.

Within the texts themselves earlier manuscript intonation marks that had served as an indicator of meaning were replaced by new, less idiosyncratic alternatives. In 1473, a German teacher in Ulm referred to these things for the first time. To understand a printed text, he said, "Watch out for the little signs," the new punctuation marks, which did away with the requirement to read aloud so as to comprehend a text and make comprehension a lot easier. The act of reading had become private.

Typography arrested linguistic drift and enriched standardized vernaculars as languages annexed dialect terminology. Thanks to print, language itself had become a vehicle of conformity and codification, paving the way for linguistic "purity."

The duplication of vernacular primers and translations contributed in other ways to nationalism because they made it possible for a "mother tongue" learned at home to be reinforced by teaching children to read the same language in print. As with the alphabet in Greece two thousand years before, in the learning years of childhood the eye would now see a standardized version of what the ear had earlier heard. And once grammar schools started giving primary instruction in vernacular instead of Latin text books, linguistic and national roots became one and the same thing.

The result was perhaps most obvious in the English Elizabethan culture and language, where through the circulation of printed books the English language rapidly standardized all across the realm. The clearest example of the way this contributed to the stability of English as a vernacular language dates from the first use of the King James Bible, introduced in all English Protestant churches in 1611 (and still there in 1970). With the help of the printed word, England was virtually a united cultural and linguistic entity by about 1600. From then on, whatever new groups, classes, or even countries might subsequently become part of England, they would be absorbed into a community already defined by print technology.

The new print-languages created unprecedented ease of domestic communication among speakers of the wide variety of accents that existed within French, English, or Spanish, who might otherwise have found it difficult or impossible to understand each other in conversation. By reading their formal common tongue on the page, they became aware of the hundreds of thousands, even millions of people in their own particular language field. In consequence, in a development that would last until the late twentieth century, they began to take pride in this new nationalist perception of themselves. There was now an "English" or "French" or "Spanish" way of thinking.

In England, the crown had been quick to realize the potential of print to induce ideological conformity and so issued a vernacular Book of Common Prayer in 1549. Among the main justifications given for its introduction were economy of production and uniformity of worship. Its creator, Cranmer, wrote in the preface: "By this ordre, the curates shal nede none other bookes for their publique service, but this boke and the Bible: by the meanes wherof, the people shall not be at so great charge for bookes, as in tyme past they have been." Cranmer's prayer book brought all the texts for all rites of public worship within the covers of a single book. He added: "And where heretofore, there hath been great diversitie in saying and synging in churches within this realme: some folowying Salisbury use, some Herford use, some the use of Bangor, some of Yorke, and some of Lincolne: now from hencefurth, all the whole realme shall have but one use."

Religious life was also made more nationalist. Previously, liturgical books had been produced in isolated monastic scriptoria where localized ceremonial traditions were able to develop, but now the press made possible a new kind of national uniform ritual.

Henry VIII of England also ordered the systematic standardization of grammar, spelling, and punctuation into "one absolute and uniform sort of learning." Education and religion were cast in the same conformist and vernacular mold, as the introduction to the 1542 edition of William Lily's grammar, *An Introduction of the Eyght Partes of Speche,* makes clear: "And as his maiesty purposeth to establyshe his people in one consent and harmony of pure & true religion: so his tender goodnes toward the youth & chyldhode of his realm, entedeth to have it brought up under one absolute and uniforme sorte of lernynge . . . consideryng the great encombrance and confusion of the young and tender wittes . . . by reason of the diversity of grammar rules and teachinges." In 1545 came the authorized Primer, published by Grafton, "for avoyding of the dyversitie of primer bookes that are now abroade . . . and to have one uniform ordre of al suche bokes through out all our dominions."

As a result of this kind of coordination it was possible, within months of the publication of a liturgical text, for royal parties of visitation to roam the countryside checking up on parish compliance with directions for its use.

The printing press gave the new, nationalist authorities the power to influence and direct the affairs of large sectors of their populations, even down to grass-roots level, in ways that seem remarkably modern. In the battle that followed the dissemination of his antipapal theses, Luther used the press as a propaganda weapon to put his case. Thousands of pro- and anti-Luther handbills, flysheets, broadsheets, and posters appeared everywhere. High in the Alpine passes, on his way to a meeting with the German emperor, Luther found printed notices calling for the banning and burning of his books only months after they had been published. Thanks to print, Europe was involved in the first continent-wide propaganda war in which larger segments of the population were able themselves to read the issues.

The potential of the press for extensive bureaucratic control did not escape the attention of governments. In Venice, toward the end of the sixteenth century, the first printed population census sheet appeared. Printing made administration easier by standardizing and simplifying the form, content, and distribution of public documents. In the Holy Roman Empire, came the first printed laws regarding disturbance of the peace, along with public excommunications and laws regulating amnesties, begging, peace treaties, or the watering of wine. With the aid of the press, laws could also be compiled in printed collections, and with many printed copies of case histories reference to precedent was now more efficient, reliable, and common.

Rules for social behavior were now set down, everywhere, in black and white. In the century and a half after Gutenberg, the press had rationalized laws and regulations to an unprecedented extent. In France, in mid-sixteenth century, the process of first codifying and then reforming the innumerable customs of the country was slow, but by the second half of the century the painstaking work of royal commissioners was beginning to achieve uniformity, with identical dispositions being followed in every bailiwick and the widespread adoption of the customs of the *parlement* of Paris.

In Spain, a codification of the laws of Castile in 1484 was followed by the printed *New Compilation* of 1567, which contained some 4,000 articles. Similar codes were issued for the other Iberian kingdoms. In the Netherlands in 1531, Charles V initiated a program of codification similar to that in France. Print also helped to standardize civil procedure relating to the family, as well as to property, succession, contracts, and other issues. The citizen no doubt felt safer in this new homogeneous, less arbitrary world of print-justice, even though printed regulations constrained freedom of action more than ever before.

As has been said, printing diffused power outward from the old papal center to the new nation-state periphery. It then isolated people within their new states' boundaries because it bolstered a new sense of national separatism, as commercial activity became easier to regulate and manage with the aid of printed passports, safe-conducts, mandates, invitations, legal notices, and national paperwork of all kinds. And as print encouraged standardized regulation of trade, the economies of the new nations began to grow and develop their own distinctive character.

The cheap, popular books flooding from the presses also quickly created a large new reading public, not least among them merchants, who typically knew little or no Latin. Political and religious printed propaganda could also be used to mobilize this growing, more literate, middle class. The circulation of broadsides and engravings carrying pictures of kings and princes heightened royal visibility. The effect of duplicated images and portraits of rulers, framed and hung both in great houses and peasant hovels throughout Europe, raised the mass-media creation of a public image, which had been used by the old Roman emperors, to hitherto unforeseeable new heights.

The most prolific early use of the press for these purposes was by the Hapsburg Emperor Maximilian, who, between 1489 and 1500, issued no fewer than eighty-five broadsheets, as well as numerous publications, giving his reasons for going to war, for levying taxation, and concluding treaties. He even produced the first government White Paper.

In an attempt to cultivate his reputation as a multitalented superman, Maximilian also planned a series of printed books and posters designed to glorify his family name. He commissioned Dürer and Holbein to engrave a massive, printed "Triumphal Arch" on paper, illustrating the genealogy of the Hapsburg house and filled with references to Maximilian's exploits. The finished work came in ninety-two separate sheets that fitted together as a giant wall covering over twelve feet high. Another commission, Maximilian's "Triumphal Procession," was even more grandiose, made up of a hundred and thirty-five large prints and stretching one hundred and seventy-five feet.

But even though print made propaganda and social control easier, it was a gift that worked two ways. The presses also made dissent more effective and so their use was quickly regulated through censorship, first by the Catholic church and then by every European monarch. To all authorities, the sheer volume of printed output represented a massive potential threat to social stability and conformity. In 1559, the Catholic church curtailed all vernacular Bible translation in Italy and permitted it only in those countries where the Reformation posed a danger. One measure of the way the presses were staying ahead of the censor is indicated by the fact that the new printed papal "Index of Prohibited Books," authorized by the Council of Trent in 1563, had to be republished no fewer than ten times in only thirty years.

Nothing gives a better sense of the concern among those in power regarding the opportunities for subversion by print than the 1535 panic-stricken ban, by the Catholic French king Francis I, on the printing of *any* books in his realm, upon pain of death by hanging. The reason for this desperate action was that French borders to the East were lined by Protestant states and cities, each producing a massive number of books that could all too easily be smuggled into France. The new ban attempted to ensure that *no* books would be printed in France, so that any new books discovered in the country would, by definition, be illegal.

In the Protestant community, the drive for literacy that would bring the word of God to every reader introduced other, more intimate forms of control. It was easier to modify behavior in any literate home by means of books than could ever be managed from the pulpit, for the simple fact that there were more books than clergy. The Puritans would later recognize this and introduce a strict set of printed standards and rules for household customs. Collected volumes of domestic "advice" purveyed this view in guides like the 1598 "For the Ordering of Puritan Families According to the Direction of God's Word."

By the early 1660s, the rules of Puritan church conduct were specified in printed official injunctions distributed widely enough even for lay people to be able to use in order to identify and report ecclesiastically deviant behavior among congregation or clergy.

Authorities were also able to foster among their print-ruled subjects a collective sense of national culture and identity, thanks also to the way the press had generated a new sense of history. Initially this was triggered by the early-sixteenth-century publication of the classics of ancient Rome and Greece, which had amazed and excited a growing number of Renaissance readers.

Now kings (and the axemakers who worked for them) began to look to their own history for proof of dynastic respectability. In the sixteenth century, the Englishman William Camden wrote a history called *Remaines Concerning Britaine,* compiled, he said, out of "love of country," and including the lives of kings, as well as descriptions of the country and its inhabitants, languages, names, arms, coins, clothing, high roads, towns and cities, scenery, and natural resources. In the same spirit, a group of English historians founded the Elizabethan Society of Antiquaries in 1572 to study and preserve old English manuscripts, while in 1577 William Harrison published *An Historicall Description of the Land of Britaine.*

At the same time, Italy acquired a national historian in Francesco Guicciardini, with his *History of Florence* and *History of Italy,* dealing chiefly with the diplomatic affairs of the Italian states from the invasion of Charles VII in 1494 to the election of Pope Paul III in 1534. Though Spain was unified territorially if not culturally, it, too, had a patriotic historian in Juan de Mariana, whose *Historiae de rebus Hispaniae* was written to acquaint Europe with Spanish history and then translated into the vernacular. In 1555, Olaus Magnus published a large work on the Nordic peoples in order to show the superior accomplishments of the Swedes, while the first Lithuanian history was written in the seventeenth century by a Jesuit, Albertas Vijukas Kojalavicus.

Germans delved into their past to find evidence of ancient Germanic civilizations and in 1455 were greatly excited by the rediscovery of Tacitus' *Germania* in the monastery of Hersfeld. From it they triumphantly reconstructed an ideal German type, based on the way Tacitus had contrasted the truthfulness, freedom, and simplicity of the German barbarians with the degeneracy and servility of his own countrymen. So from the testimony of a Roman himself, the Germans could assume the superiority of the German character over their contemporary Europeans.

These Germanic claims to superiority went to extremes when some claimed Adam had been a German who had spoke Alamannic, a language which must therefore have been humankind's original tongue, and which would once again be restored to its dominant position when the Empire attained world control and established the true *pax Germanica.* When the humanist Heinrich Babel was crowned laureate by Maximilian at Innsbruck in 1501, he claimed in his address that the Germans had conquered practically the whole Earth and had subjugated many peoples.

More practically, from the point of view of kings and princes, the new histories strengthened the sense of separateness in the new nation states and, especially in the Protestant countries, aided monarchs in their attempts to move their administrations away from papal control. This was to be made easier by a chain of events, set in motion thanks to Rome's original introduction of printed vernacular Bibles, that would finally lead to the diminishing of Catholic

Christendom, with unexpected and shocking effect. In this case, the outcome would attack the core of religious belief itself and confer on secular authorities a new gift with which to cut and control the world.

The sequence was triggered in 1545, when Rome convened an ecumenical council in the northern Italian city of Trent to discuss measures to combat Luther. As part of general measures to standardize worship, the Council authorized the publication of approved versions of all liturgical Catholic texts. In Antwerp (at the time under the control of the Catholic King of Spain, Philip II) was Christopher Plantin, a printer who ran the biggest publishing house in Europe. Plantin had an extensive network of sales agents and offices, from Norway to North Africa, through which he sold books and ran a profitable sideline in the import-export of plums, wines, French lingerie, fancy leather goods, mirrors, and scales.

Plantin had founded his print shop in Antwerp in 1555, which, at the height of his success, housed twenty-two presses, nearly two hundred staff, and an organization five times bigger than that of his nearest rival. Plantin's new printing house epitomized the new entities springing up all over Europe, bringing together an entirely new mix of intellectual and commercial disciplines. This in itself was revolutionary, since before printing these separate areas of specialist knowledge would have had no reason to interact. In Plantin's shop, university professors and ex-abbots acted as proofreaders and text editors, scholars of all subjects checked text for factual accuracy, artists prepared woodcuts and engravings, craftsmen printed or advised on books relating to their own area of expertise, and merchants became involved as financial backers.

Plantin and his fellow printers were also the first real capitalists, raising money to support their ventures, giving their financial backers a share of the profits, developing production schedules, linking sales to marketing, organizing labor, and negotiating with strikers.

The Plantin establishment and others like it were a mixture of sweatshop, boarding house, and research institute.

Plantin's print house would change history, because it would enable the axemakers to produce the greatest force for change so far, modifying the nature of knowledge itself, and suddenly widening the gulf between those with specialist knowledge and those without.

This happened because in 1566 Plantin wrote to the secretary of Philip II of Spain to suggest an entirely new kind of Bible. It would, he argued, fulfil the Council of Trent's wish to consolidate the power of Rome and better control the laity. More important, it would add luster to the reputation of Philip himself. To be printed in all the biblical languages (Latin, Greek, Hebrew, Syriac, and Aramaic), the new Bible would be based on the new, print-generated, analytical approach to textual criticism.

Over the centuries, both classical and biblical manuscript texts had accrued large amounts of commentary and explanatory material, generally written in the margins. When these texts were first printed, it had become customary for the additions to be incorporated and the whole text submitted to detailed examination so as to detect repetition, textual mistakes, and correct all-too-frequent manuscript copying errors. During this work, editors developed new ways of thinking about the knowledge in manuscripts, using skills in textual, factual, and grammatical analysis that had not been considered necessary before and which Plantin would now apply to the production of his new Bible.

In 1568, he received permission to begin the work, and five French and Flemish scholars gathered under the personal supervision of King Philip's theological adviser Benito Arias Montano, who was to run the enterprise eleven hours every day for four years. In 1572, Plantin finally printed 1,212 copies of the eight-volume work, by this time known as the "Royal" Bible, the first five volumes of which carried the biblical text set out in the five languages. The last three volumes contained a new type of knowledge only made possible by printing. These were volumes of additional material, containing commentaries and information relative to the biblical text and based on compendia of the most recent scholarly discoveries.

The appendices, edited by Montano, held vast amounts of data on everything from biblical genealogies to maps of the Holy Land, notes on the Hebrew idiom and on the origins of the language, plans of the Temple in Jerusalem, Jewish antiques, histories of the tribes of Israel, and essays on biblical coins, weights, and measures. There were also Aramaic, Syriac, Greek, and Hebrew dictionaries and grammars, variant readings of text, discussions on the meaning of the terminology used, indices, and no fewer than eighteen treatises on archaeological and philosophical matters.

The idea caught on and other Bible appendices began to appear all over Europe, often accompanied by woodcut and engraved illustrations of Holy Land atlases, designed by well-known mapmakers, as well as plans of the biblical cities drawn by famous artists. The material was organized alphabetically, and great care was taken to rationalize, codify, and catalogue everything.

The scholars, cartographers, lexicographers, and others who came to lend their expertise to the printers had gained their knowledge and a new analytical approach to information from their earlier, editorial work carried out while collating print versions of classical works on botany, herbals, zoology, petrology, medicine, and anatomy. People who had worked on these texts and on the polyglot Bibles now began to express their talents in other, more innovative ways. Montano, for instance, went on to write a history of the world that drew on the latest archaeological material and that provided so much new, uncensored information that he eventually fell victim to the Spanish Inquisition.

When the biblical scholars finished their work on the appendices, they turned, like Montano, to other things and, as they did so, triggered an extraordinary "knowledge fallout" that was to affect almost every aspect of sixteenth-century European life and that would help to shape the modern world. The new knowledge-specialists set up intellectual book networks across Europe, exchanging anything from maps to data on instruments, flower bulbs, plant seeds, and rare stones.

The astronomer Johann Kepler's early work on biblical references to the heavens (and the bibliographical skills this had taught him) gave him the background for his epoch-making publications on planetary dynamics. Compilers of biblical dictionaries went on to produce modern language grammars and dictionaries for use by merchants and traveling traders. In 1617, for instance, a *Guide to Eleven Languages* appeared in England.

Perhaps the only other printed works with a market potential as big as that of the biblical appendices were the almanacs. These also attracted the skills of a wide variety of the new "experts." Almanacs had existed in limited form before printing, but from the sixteenth century on they appeared all over Europe in massive numbers and radically increased the amount of data in circulation. They were, by this time, aimed at a more general readership, and the texts included seasonal compilations of useful information of all kinds from lunar, solar, and tide tables for sailors to agricultural data and animal husbandry for farmers, prognostications and astrological tables for the credulous, calendars for the merchant and the devout, notes on childbirth for midwives, commercial arithmetic and commodity prices for traders, and weather forecasts for all.

By 1600, English almanacs were already being standardized, usually including a section on calendrical and legal data, as well as material on the seasons, with relevant medical topics and notes on agriculture. Often a section would be added, giving lists of trade fairs and their locations, and one edition even added a blank diary page opposite each month's entry for those readers who were attending the fairs. In early-seventeenth-century Europe, these almanacs were selling 400,000 copies a year.

As time passed, the data contained in the almanacs in turn encouraged the growth of more specialist disciplines, each of which needed its own almanac: the seaman's calendar, the weaver's almanac, the constable's almanac, the farmer's almanac, and so on. Each publication standardized the specialist data and helped to institutionalize agreed practices and rules for entry to and work within the craft. In trades and occupations, the almanacs also helped build new trading networks by providing uniform tables with which to compute the cost of goods and payment of wages, as well as weights-and-measures conversion tables and distances between markets.

As the interest in classical science and technology increased with the publication of specialist works in print, compilations of classical knowledge proliferated. Doctors benefited from them in 1543 with the Belgian anatomist Vesalius' work, followed soon afterwards by works on contagious disease, pathology, reproduction, and surgery. Merchants and mathematicians had Pacioli's textbook on algebra and geometry as early as 1494. Architects, astronomers, and surveyors had their first printed edition of Euclid, in Basel in 1533, the same year as Regiomontanus' work on trigonometry and Frisius' on triangulation. Shipbuilders read the first print edition of Archimedes in 1544, and in the eight years following 1551 nine major studies were published on animals, birds, and marine life.

There were also books on every subject useful to a rapidly growing community of axemakers working in the fields of architecture, ballistics, surveying, magnetism, machinery, cosmology, astronomy, navigation, warfare, fortification, geology, assaying and metallurgy, dyeing, and textiles. This outpouring of regularly updated literature disseminated standardized technical information for the first time and encouraged the rapid development of new specialist skills. These textbooks and manuals began to undermine the position of old people. Where once youth had sat at the feet of their elders to learn from skills accumulated over decades of experience, these could now be learned by opening a book.

With each new publication, knowledge became ever more fragmented and esoteric. After the early editions had summed up the work of the classical authorities, specialists now knew enough to try and judge from their own experience. Anatomists, for instance, opened bodies and saw for themselves the errors of the ancients regarding the position of the organs and circulatory systems.

Perhaps the most innovative work of this type was that of the botanists at the Lutheran University of Wittenberg. Luther himself was a keen naturalist and lover of plants (his emblem was a rose), and he wanted natural knowledge to become the property of ordinary people. At Wittenberg, the botanists benefited from the presence of a medical school and the inclusion of classical botanical authors in the arts curriculum. The reason for this particular interest in botany was that most contemporary medicines were made from herbs. So in 1529, Caspar Cruciger, Wittenberg professor of theology, established two botanical gardens outside town because he felt that sooner or later plants would provide the cure to all diseases.

Valerius Cordus left Wittenberg on a botanical trip in 1542 that took him throughout Germany and Italy, and he produced *A History of Plants,* printed in 1561. Rauwolf, another Lutheran, traveled from Augsburg to the near East and collected a herbarium of 843 plants, still preserved at the Dutch university of Leyden. The Dutchman Willem Quackelbeen journeyed through Turkey and brought back the horse-chestnut, lilac, and tulip. Before Gutenberg's press the ancient classical authorities had identified six hundred plants, but thanks to print, by 1643 the total had risen to over six thousand.

Everywhere the gift of print generated a new way of thinking about the world. There were clearly things which the classical authors had not known, things not yet discovered, and ways to improve that which was already known. So print engendered a fever for novelty. The presses gave Europeans a powerful desire for progress and change because they made people aware of history and now offered new knowledge with almost every new edition. After mid-sixteenth century, most of the titles of specialist books printed (there were, it must be remembered, almost no nonspecialist books) included the word that would be associated with axemakers from then on: "new" (*New Science, A New Theater of Machines, A New Tool,* and so on).

The problem for the authorities was to what extent and in what form all this novelty could be safely disseminated without causing disruption. Since the early years of printing, the proliferation of data had spurred institutional attempts to make certain that literacy would not prove to be socially destabilizing and to ensure this by using the book as an agent of social control, limiting the amount and kind of new knowledge offered to the general public.

Textbooks, made available in large numbers by the presses, standardized knowledge and made it ideologically acceptable, but even then not everybody approved. Books were seen by some as "the infatuation of the people," and as early as 1498, in Mainz, it had been said that "everybody now wants to read and write." Education took on new significance in ensuring the proper management of knowledge. The number of older Catholic schools was soon overtaken by those of the new Protestant religion, because Luther was urgently concerned with the education and indoctrination of the young. Obsessed with the need to create a totally ordered, hierarchical society of literate and docile believers, Luther structured education into a classified and graded process that submitted learners to standard examinations that would identify their level of ability and reveal deviance or ignorance.

The process of streaming helped to single out those with the potential to be admitted to positions of authority. New pedagogic experts emerged to control and administer the new indoctrination process. At Lutheran urging, curricula became official, teacher training was state-controlled, approved texts were printed, and use of the vernacular rather than Latin ensured that the new regimentation would reach down to the lowest stratum of society.

In Northern Germany, the ability to read was made a prerequisite of confirmation, which was itself the prerequisite of marriage and, in this way, those who would marry had first to pass through the system and be approved. Between 1530 and 1600, more than a hundred ordinances dealing with education were promulgated in Germany alone. In the Duchy of Württemberg, for instance, a vernacular elementary and Latin secondary school system was established in every town and village to train boys for employment in church and government, with five-grade schools, a uniform curriculum, and standard, printed textbooks studied in preparation for a single state-wide examination. The Danes, Swedes, and Dutch soon copied the German example.

The Catholic church responded to this threat by setting up Ignatius Loyola's great Jesuit colleges, the first of which was opened at Coimbra, Portugal, in 1542. The emphasis here was on teaching Latin and training priests, but the Jesuits became increasingly involved in public education, and in 1546 began admitting secular externs to their colleges. The Jesuit stress on uniformity and high standards of achievement—in the context of continent-wide, centralized administration—produced a pedagogic system that was, in terms of social control, far in advance of anything else in Europe. The course material was extremely formalized and taught under strict theological supervision, while students were monitored at all times by teachers, prefects, and rectors to make sure they conformed.

Luther and Loyola had, in different ways, established the educational system as the principal agent of the different belief systems, but by the time of the establishment of the first seventeenth-century Royal Societies of Knowledge, another key figure had emerged in education. He was a Czech called Amos Komensky (Latin name: Comenius), and his apparently democratic aim was "to teach all things to all men." Komensky's *Great Didactic* was the most important educational treatise of the century, and in it he detailed "a method for teaching children the sum of all knowledge," for which gift he has been called the Father of Pedagogy.

The focus of his interest was what concerned educators most: the need, in a time of rapidly growing trade and commerce, to use education as a tool for inculcating "useful" knowledge. Children in school were now to be given tools so that they might experience work and make their choice of vocation early. However, Komensky stressed the potential of education to control and predict human behavior, "for there will be no ground for dissenting, when all men have the same truths clearly presented to their eyes."

English grammar schools, French lycées, and German gymnasia

followed the vocational lead of Komensky because their merchants were eager to prepare the young for the new capitalism by having them taught in establishments free of direct church control. The new commercial schools taught reading, writing, and arithmetic, the essentials for success in the expanding economy.

So, as print made possible the transmission of a growing body of information without risk of loss or corruption, the new educational institutions in turn helped to train people for new, specialist organizations that could profit from "useful" knowledge. It was only a matter of time before bureaucracies, themselves controlled by the state, were set up to sift out the socially most applicable data and regulate its use in the form of coteries, which would become known as "professions." Their principal aim would be, as always with specialists, to defend and maintain the exclusive nature of their particular body of knowledge and to support their royal patrons and the social order.

In 1518, the English king Henry VIII, seeing the wisdom of bringing these new knowledge-makers under government control, founded the Royal College of Medicine in London. The College had powers to prosecute malpractice and to issue its own statutes, pass its own ordinances, organize meetings, and grant licenses. In 1540, the English surgeons set up a new Company to regulate the new profession of surgery, entry to which now required an examination on theory and practice, the subject matter of which was prescribed and standardized in great detail. In the same year, the Royal College of Physicians was established with similar powers, and in 1617 the Society of Apothecaries was founded, with three grades of membership and its own Hall and herb garden.

As the professions established themselves in print, their members increasingly used printed books to communicate with each other in language that became more and more incomprehensible to all but members of the profession. However, this esoteric community of reading and writing axemaker specialists would be shaken by a number of events that appeared to threaten the very credibility of the printed knowledge on which their new power rested. The effect would be to generate a radically different view of knowledge, where it could be found, and what could be done with it.

The event that brought this momentous change about was so extraordinary that when news of it became generally known (about thirty years after it happened), the majority of people simply didn't believe it. It was the discovery of America.

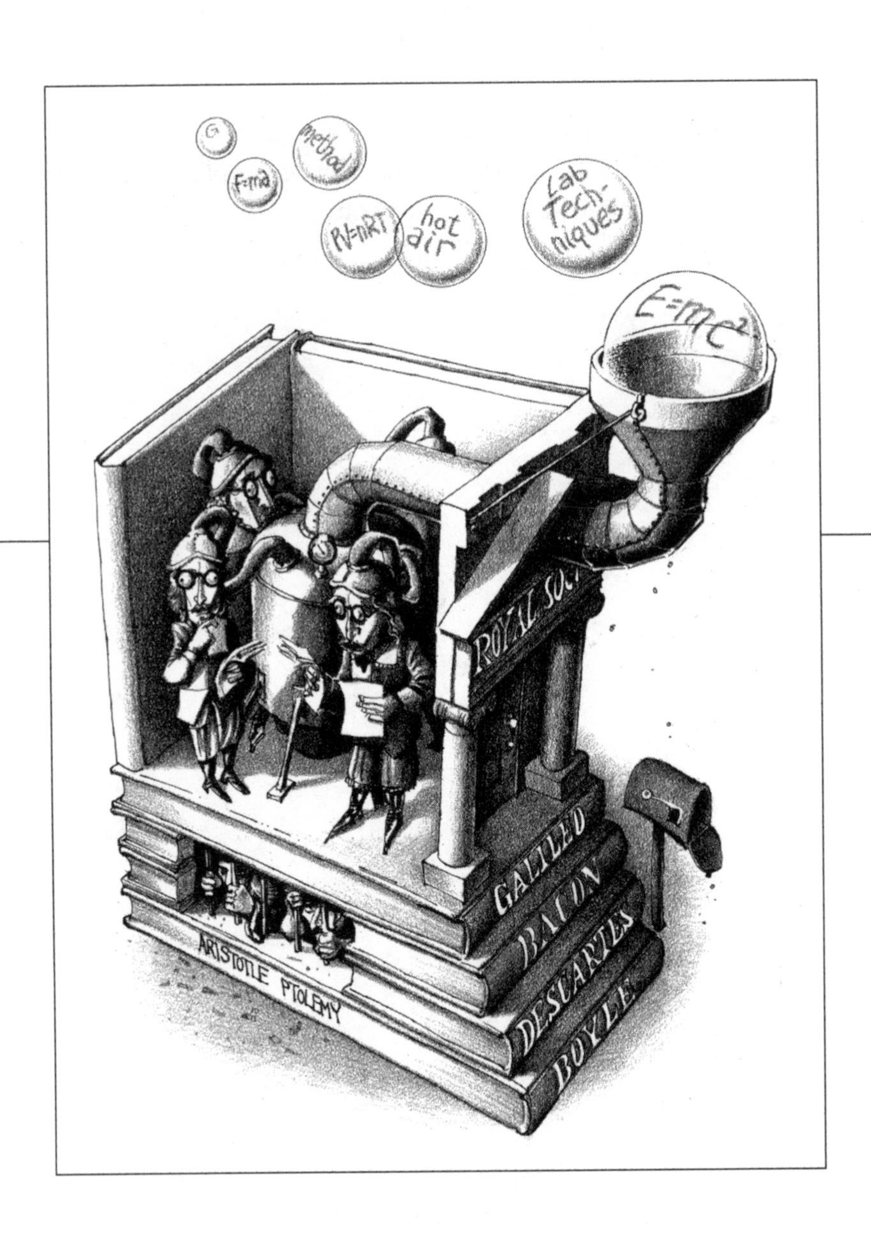

G
F=ma
method
PV=nRT
hot air
Lab Tech-niques
$E=mc^2$
ROYAL SOC
GALILEO
BACON
DESCARTES
BOYLE
ARISTOTLE
PTOLEMY

Chapter 6

NEW WORLDS

On July 22, 1502, the Italian manager of the Spanish branch of a Florentine shipping agency landed in Lisbon at the end of a voyage to Brazil. It was his third transatlantic crossing and it made him famous because in 1507 a version of his name was printed in a small insert on a map drawn by the French cartographer Martin Waldseemüller, showing where the sailor, Amerigo Vespucci, had gone.

Waldseemüller's map took Europe by storm because it showed the world as Vespucci had seen it, with an extra continent in the middle of the Atlantic and another ocean beyond, separating it from Asia. Columbus had never expressed this view and, in any case, little or nothing had been heard from him since 1497.

In 1505, Vespucci recorded his journeys in a letter he entitled "The New World," which by 1527 had been published in twenty-three Latin editions and thirty-seven vernacular versions in every major European language. By then the continent was already being referred to by the name that Waldseemüller had coined from Vespucci's first name: "America." And the discovery of America was to make quite a splash.

From the time of the first flint tool, axemakers' gifts had given the institutions of leadership the means to reshape the world. Each time they did, entirely new structures and systems appeared in the form of carpentered hunter-gatherer shelters, law with which to control the cities of Mesopotamia, Greek logic to enforce conformity on the investigation of natural processes, the medieval trick of "reproducing the phenomenon," and the new regulated print professions. But now an entirely new kind of knowledge was to emerge as it became clear that the world was not all it seemed.

It is ironic that the new knowledge would be triggered by an event that surprised even the axemakers, with a flood of knowledge they had not themselves manufactured and which came from a source they had not even identified. Their response to the problem this created took the form of a gift that would bring to the community material benefits far beyond anything that had been offered before and at the same time would remove specialist knowledge entirely from public gaze and place it in new, artificial worlds. We call it "science."

Vespucci's accounts of the New World destabilized European society because it began a process that would eventually call into question the validity of the premises on which all social power had rested up to then. European core ideological belief was that the Earth was the center of the universe; that everything in the heavens and on Earth had its proper, God-ordained place; that there were only three continents; above all that, according to the Bible, all these matters were fixed by God at Creation and were therefore incapable of being changed.

It looked as if the discovery of America, which the Bible had not predicted, would sweep all this away, and with it the entire social structure built on two thousand years of theological and philosophical authority. The medieval system of arguing the way to the truth was a failure because it had not prepared people for the possibility of an extra continent.

An even greater problem came from the awkward fact that America was inhabited by primitive people in a state of natural existence, apparently without knowledge of politics or history or Christianity, and yet there they were, without all this knowledge, surviving quite happily in voluntary, organized, and functioning societies. This realization triggered the rapid spread in Europe of the radical concept of "free association," a social principle that might (it was said) prove to be better than the old European social forms whose existence depended on submission to authority. Free association proposed that communities of people should come together of their own free will and voluntarily agree on their laws, just as "primitive" American tribes appeared to do. This concept was pretty shocking in the sixteenth century, but a hundred years later it would influence the thought of John Locke and through him the founders of the French and American republics.

America was apparently untouched by European or any other influence, so at least it offered a unique opportunity for the study of God's work free of the constraints of classical thinking. But anybody who might have wanted to try their hand would be in dangerous philosophical territory since no official yardstick existed to categorize and control the incorporation of this new data into society. So first and foremost there was urgent need on the part of the institutions of church and state for a new definition of a "fact" that would be fail-safe, so that new information could be made socially acceptable.

The search for this new definition would radically redefine the concept of knowledge itself and bring the new tool for change that would enable axemakers to create whole new worlds. But for the authorities it was first and foremost essential to develop a way to control how new data was to be collected and to decide who would then be allowed to know it. New World amazements, like pineapples, potatoes, turkeys, and cactus, caused a stupefied review of the techniques used in the study of natural history. The discovery of unknown species proved the superiority of direct observation of nature and pulled the rug out from under the previous, uncritical use of classical definitions. Initially, attempts were made to keep things within the limitations of the old categories, but their inadequacies were soon revealed by terminology of the type that was used to describe the newly discovered tapir: "part-bull, part-elephant, part-horse." The English term "pine-apple" and the French *pomme de terre* (earth-apple = potato) also reflect this early attempt to make old descriptions fit new things.

But the immediate urgency was to categorize and name everything. Once things had names, it was thought, they could be controlled. The first European botanical garden set up to receive new species for classification was laid out in 1545 at Padua, Italy, and was modeled on Cortez' descriptions of the huge gardens of Montezuma. By 1577, Juan de Ovando, president of the Spanish Indes, had organized an information-gathering bureaucracy that distributed printed questionnaires to everybody coming back across the Atlantic. When specialist archives were set up in Seville a few years later, the administrative structure for managing and ordering American data was in place.

But as more and more travelers returned from the new continent, they brought with them questions that undermined authority of all kinds. How could these wild, naked Amerinds be descendants of Adam and Eve? If Ptolemy, the unquestioned Alexandrian classical authority on geographical matters and on whose maps all contemporary cartography was based, had not known of America, how accurate was the rest of his data? If the supreme Aristotle had been wrong about the number of continents (he had said there were three), was his entire method of classifying nature, the bedrock of Western intellectual life, to be trusted?

But worse was to come. At the same time as doubts were being raised about the nature of the world, similar questions were also being posed about the nature of the cosmos. Official contemporary cosmological thinking followed the church's Aristotelian description of the universe as a series of invisible, turning spheres, made of some unearthly material, on each of which rode the Sun, the Moon, the planets, and the stars. At the center of everything stood an immobile Earth.

This cosmic system had for centuries provided the basis for social management because it provided the calendar of liturgical holy days that the faithful were duty-bound to observe, since failure to do so would jeopardize the individual's chances of salvation. Unfortunately, astronomical calculation had remained virtually unchanged since the first century, when Ptolemy had modified the Aristotelian model, introducing the concept of some planets turning on smaller spheres attached to their main sphere, so as to account for the way that bodies like Mars seemed sometimes to go backwards. By the sixteenth century, the system had been modified to include no fewer than ninety of these "epicycles" and was hopelessly inaccurate. The exact calculation of Easter from the relative positions of the Sun and Moon was now impossible. Ptolemy's system had lost the church's prime festival.

But it was theologically vital to find Easter again, so Rome asked one of its astronomers, a Polish canon called Kopernik, to solve the problem. Copernicus (the Latin name by which he is more commonly known) did so, but in 1543 was obliged to come up with an alternate solar system arrangement that put the Sun at the center of a solar system in which the Earth was just one of several planets.

The implication of this new system was philosophically mind-boggling because it denied the centuries-old claim that humankind held a special position, central in the universe, as befitted a creature made in God's image. This undermined the supreme authority of the church, which rested on just such an assertion. For a while, Rome succeeded in fudging the issue by describing Copernicus' gift as a "mathematical fiction" there merely to "save the appearances" with a purely theoretical construct. In any case, God knew that Copernicus was wrong.

But throughout the sixteenth century traditional cosmology was gradually undermined by more new astronomical data. Appearances by comets and supernovae again challenged the Aristotelian view that the sky was incorruptible and unchanging. Also, the special nature of the material from which the heavens were made was supposed to mean that the sky was the only place where circular movement occurred. According to Aristotle, earthly movement could only ever be rectilinear. Objects thrown (or fired from cannon) were presumed to move in a straight line and then, when they ran out of "impetus," to fall directly to the ground.

And then the Italian artillery specialist Niccolò Fontana, known as Tartaglia ("the stammerer"), upset everything because when he tried to find a way for his guns to achieve maximum range he discovered that all ballistic trajectories were actually curved. This revelation of a flaw in the Aristotelian scheme gradually brought about the realization that phenomena had to be described in a more reliable way, with the aid of measurement. The attempt to do this would bring about an entirely new technique for deriving data from and exchanging data with the natural world, because it would introduce nonreligious explanations for the world, based on an individual's observation. But uncontrolled, personal, quantitative assessment of the world was the last thing the church wanted because it encouraged questions.

The first exponent of this world-changing new technique was the Italian Professor of Mathematics, Galileo Galilei. In 1603, in Padua, Galileo tried an entirely different approach to the study of natural phenomena, in which he first worked out the answer to a problem mathematically and then looked for proof through experimentation.

He used the new method first of all on the problem of the acceleration of falling bodies. Aristotle had said each object fell because it sought its natural position on the ground and that it moved faster as it did so because of a sense of happiness at nearing its goal. Galileo proposed a radical alternative (that also explained Tartaglia's curved trajectories) according to which all bodies fell in the same way and at the same rate, because they were all obeying a common law of nature that could be mathematically derived and experimentally proved.

Galileo rolled a ball down an inclined plane marked off with gut-string frets and then, using a pendulum, measured the distance the ball went in equal time periods. The experiment revealed that the distance of the ball at any time during its roll from rest was always the square of the elapsed roll-time. In this way, Galileo had shown that he was able to proceed from a mathematical abstraction to its proof in experiment and then to a generalized statement of the consequent law governing falling bodies. Above all, he had shown that it was possible to model nature in mathematics, analyze any problem in terms of its basic principles, and generate certain knowledge that was universally applicable.

At the same time, an aristocratic English lawyer named Francis

Bacon was approaching the issue of certainty in knowledge from a different, though complementary, direction. In 1620, he proposed nothing less than an entirely new approach to the problem of generating information in a major work called *The New System* (because it replaced Aristotle's *Organon,* the "old" system used until then for analytical thinking).

Bacon believed that the Copernican system, the discovery of America, and the new data coming in from around the world had all created a general crisis of knowledge. He wrote: "The mind of man . . . is like an enchanted glass, full of superstition and impostiture . . ." America and the new cosmology left no solid ground for authority, there was also no longer one united Christendom, Galileo's discovery of mountains on the Moon had disproved Aristotle, and Copernicus undermined the fundamental stability of the universal model together with the social structure based on it. The English poet and churchman John Donne expressed the near panic caused by the crisis: "The new philosophy calls all in doubt."

Confusion reigned because nobody knew what was now officially approved knowledge. In a world where the punishment for deviance could be death, this was a serious matter. For Bacon, the crisis was so fundamental it signaled the beginning of a new era that needed entirely new modes of thought. There could be no compromise with failed Aristotelian systems or modifications of previous forms like those of Ptolemy. The old scholastic approach had brought about a general decay of learning. In dusty university halls, scholars argued with premises now rendered irrelevant to an understanding of the dawning new world and impotent to address, analyze, and answer the new questions facing the old world of traditional Aristotelian and Ptolemaic thinking. Nobody was addressing the problem. Urgent action was needed if social stability were to be preserved.

The flood of new American, Copernican, and Galilean data necessitated radically new rules for collecting data and, above all, for putting it in some kind of overall scheme. New principles were needed, as well as new forms of argument, new aims for knowledge, a new ethic, and a new means to channel and manage the flood of discovery, which seemed to be about to overwhelm the social and intellectual order. Bacon felt the situation was so desperate that "the entire work of understanding [should] be commenced afresh."

His "new tool" grounded knowledge firmly in observation and experience because Bacon saw that the old scholastic method of judging a theory by how well it could be argued failed when faced with the unprecedented data coming from newly discovered parts of the world. Only the exhaustive gathering and classification of information would bring the kind of certainty that would maintain social stability because it would reveal, in a new way and with new kinds of evidence, the orderliness of God's creation and the regularity of the workings of nature and society.

Bacon also felt that intellectual understanding was not only the privilege of a few elite scholastics, because, after all, the empirical knowledge of ordinary craftsmen had led to the great discoveries of gunpowder and clockwork. The best results would therefore be obtained by opening up investigation to the broadest spectrum of experience. In spite of appearances, though, Bacon was not in favor of unlimited access to knowledge, preferring that facts be made available only if they were appropriate to the user's social status.

Bacon felt that it was part of God's plan that "the opening of the world by navigators and commerce and the further discovering of knowledge should meet." Facing the urgent need for social stability in a period of intellectual and theological turmoil, when the church and its servants, the secular authorities, were under attack and unsure of their ground, the primary goal should be to get hold of information that was utilitarian, life-improving, but above all reliable. So Bacon's new system would offer the new data to all potential users by communicating it in the clearest possible language to as many "legitimately qualified" people as possible.

Bacon also set the parameters for the increasing contact between Europe and the newly discovered lands and set the stage for the exploitation of those countries: "Let a man only consider what difference there is between the life of men in the most civilized province of Europe and in the wildest and most barbarous district of

New India. . . . This difference comes not from soil, not from climate, not from race, but from the arts." (By "arts" he meant new, Baconian-style knowledge.)

Bacon's new data-management process had four essential components, and they would bring into being our modern view of knowledge: finding, judging, recording, and communicating, so as "to detect and bring to light things never done." If such "things" turned out to be truly new, for Bacon it was all the more important for them to be viewed in the cool light of "objectivity," a new term that would be the watchword of the new axemakers. In a world reeling from the effect of the new and unforeseen, opinion, personal tendencies, and the evidence of the senses were now considered too subjective to be reliable. Wherever necessary, investigators should protect themselves against possible error through the use of "objective" instruments that would correct any deficiencies in human perception.

Regulation of thought by approved methods like this would make control and mastery of nature easier and, most important, it would also make easier the enforcement of conformity. With Galileo's experimentation and Bacon's system for managing data emerged the beginnings of a new definition of knowledge, of the role that theory should play, and of the "objectifying" capacity of mathematics to quantify phenomena. This left only the question of a technique for evaluating raw data without fear of error.

Even as Bacon was formulating his new system, in a small town in Bavaria a snowstorm kept a French military engineer in his lodgings for a whole day and night during which, he said, he had formulated the concept that was to solve the problem of evaluation. His method for doing so would also give specialists a powerful new gift to help them in manufacturing knowledge. In 1637, after much rethinking, the engineer, René Descartes, published his new concept in a book called *Discourse on Method,* in which he set out the rules for seeking certainty in an uncertain world.

The secret lay in what he called "methodical doubt," by which everything except self-evident truths were to be questioned until they had proved themselves to be true (and for Descartes, everything, especially the evidence of the senses, was to be doubted in the absence of any "evident truth"). Descartes' method provided the supreme cut-and-control approach to the world in the form of a technique known as "reductionism." In an echo of the medieval resolution-and-composition technique, the method called for a problem to be divided up into its smallest parts so that it could more easily be understood and then solved. All reductionist thinking should proceed from the simple to the complex and all statements about the world should be expressed only in nonmetaphysical terms: size, shape, and movement.

Descartes' overriding concern (that knowledge be used to bring social order) is revealed by his insistence on a "purification urge" that would put the world into firm, clearly articulated categories, admitting of no ambiguity or dissonance and controlling the evaluation of any experience in advance. According to the *Discourse,* anything that could not be previously categorized would not be studied. With Descartes' reductionist axe the selective and exclusive process of human perception, originally modified by language and the alphabet millennia earlier, was now even more constrained. Technology would soon be available to turn those constraints into forms that would render axemaker activity even more incomprehensible.

The metaphors that Descartes used suggest more an obsession with conformity than a desire for innovation. Descartes demanded a widespread "purge of the mind," an overturning of the "baskets of rotten fruit." His aim, like Bacon's, was to start afresh in a universe (he intimated) clear of boundaries and preordained structures, ready to be "put in order." Like others of his time, he feared the new voices of relativism like Montesquieu, a French essayist and political commentator, whose *Persian Letters* (written for safety's sake as if from the Persian ambassador in France to a friend back home) questioned whether or not the European model was any more certain or valid than that of the "savage" in America. Montesquieu's *Persian* views mocked the accepted European values and satirized the French authorities' absolutist position.

Galileo's mathematical proofs, Bacon's empiricism, and Descartes' methodical doubt produced a new investigative technique so powerful it would enable axemakers to reach unprecedented levels of esoteric specialization and control. The first widespread social effect of the new reductionist way of thinking was to generate hierarchies that would regulate how to apply the method and how to manage the flood of data from all over the world, which the application of scientific method would trigger.

These hierarchies were known as "academies for the propagation of knowledge," and in 1657 the first, the Accademia del Cimento, opened in Rome, with its motto: "Test and test again." The example was soon followed by similar gatherings all over Europe, the most exclusive of which would be in France, where membership of the "society" was limited to sixteen.

The first permanent, official body successfully to organize and manage the manufacture of knowledge (which the new technique made possible) was inspired by Bacon's thinking and started as a group of young men calling themselves the "Experimental Philosophy Club." They met regularly at Wadham College, Oxford, to talk about key contemporary discoveries, like the circulation of the blood and Copernican astronomy.

The discovery of America, the new cosmology of Galileo, and the dangerous relativism preached by radicals like Giordano Bruno (the Italian cleric who had been burned in 1600 for suggesting the universe might be infinite and contain other inhabited planets like the Earth), as well as the divisions within organized religion between Catholic and Protestant and the avalanche of new information coming in from around the world, all made the new "experimenters" keen to find a way to manage knowledge that would above all be politically correct.

While the Oxford group wanted to move as fast as possible away from the old authoritarian views of classical science, they were also concerned to place the investigation of nature on a basis that would prove easier to control. The country was only recently back in the hands of Church of England royalists and a restored monarchy after the years of Republican Commonwealth under Cromwell. But radical Puritanism was still alive, kicking, and dangerous. It was only a few years since the end of the Commonwealth, when Puritan political radicals had taken over the government after winning the Civil War and beheading the Catholic Charles I. In a once-again High Anglican Church kingdom, only a few years after the end of the Civil War, this meant keeping it out of the hands of the Protestant political and philosophical radicals, who had nearly succeeded in turning England into a Republic under Oliver Cromwell.

The Wadham group proposed the creation of a "College for the promotion of physico-mathematical learning," and in 1662 it was taken under the establishment wing when founded by Charles II as the "Royal Society for the Improvement of Natural Knowledge." The Royal Society (which still exists today) would act as a model for axemakers all over Europe for the next century and beyond, stimulating and directing the manufacture of the new knowledge that would one day lay the foundations of the industrial world. For the moment, however, the key purpose of the new Society, and others like it that followed throughout Europe, was to defend the institutional status quo by marshaling new techniques (and the new knowledge they generated) in defense against what was declared to be "atheism," the catch-phrase for anti-Establishment behavior.

Above all, there was to be no conflict between science and state religion, because the physical order of nature should be reflected in a disciplined society: the more nature was understood by science, the more the social structure could be maintained. Society members were for the most part members of the state church, the most powerful of them being Latitudinarians, a subsect within the Anglicanism that controlled most of the high ecclesiastical offices in the country, including that of the Archbishop of Canterbury.

The Royal Society's declared aim was the "controlling of matter" for use by a community, in which each individual would "attend the duties of that particular station of life, whatsoever it be, wherein providence has at present placed him." In pursuing the orderly laws of nature, science would function as an essential tool to encourage social conformity.

In his history of the Royal Society in 1667, Thomas Sprat explained the value of the new experiments to encourage such conformity:

> Transgression of the *Law* is *Idolatry:* The *reason* of mens contemning all *Jurisdiction* and *Power,* proceeds from their Idolizing their own *Wit.* They make their own Prudence omnipotent; they suppose themselves *infallible;* they set up their own *Opinions* and worship them. But this vain *Idolatry* will inevitably fall before *Experimental Knowledge;* which as it is an *enemy* to all manner of false *Superstitions,* so especially to that of mens *adoring themselves* and *their own Fancies.*

Galileo had already shown that "objective" experiments could bring new certainty in knowledge, and this technique would now be used to secure and legitimize church and state.

The Royal Society's motto was (loosely translated): "Take nobody's word for it," and it enthusiastically embraced Descartes' methodical doubt. Members realized that the application of the new scientific method was likely to inundate them with "experimental evidence" from every quarter, so they introduced regulations to make sure the data would be generated in standardized form. The aim of the Society was not to indulge in dangerous, old-fashioned metaphysical discussions, but to gather first-hand, cut-and-control data and concern itself with what it referred to as "matters of fact."

Much of the first-hand information that the Society received from "reporters" (the original use of this term) around the world came from sailors, merchants, and military personnel, as well as from English travelers or foreign observers of every kind. The Society categorized and classified the incoming data through a committee for correspondence and arranged for the publication, starting on March 6, 1665, of a journal describing the Society's experimental activities.

The new journal, *Philosophical Transactions of the Royal Society* (still published today), was the first scientific periodical, and it set the style for all the others that followed. Its rules of acceptance required that reports be submitted in a format approved by the Society, which would make it easier for editors to spot politically or theologically deviant thinking. The inaugural edition included reports on telescopes in Rome, observations of a spot on Jupiter, a French comet prediction, a new lead ore from Germany, a letter about whale fishing in Bermuda, and an evaluation of sea-going pendulum watches.

In order for evidence to be "objectively" tested, new procedures were instituted by the Society's leading member, the Irish aristocrat Robert Boyle. Boyle's view was that science ought to reveal God's grand design and strengthen orthodoxy, so he reasoned that the best way to arrive at an "objective" assessment of evidence was to have every experiment repeated before a number of Society members in an act called "multiple witnessing."

Only when something was collectively observed in this way, by the consensus, could it be safely accepted as a matter of fact. A special vocabulary and a standard way of recording data were also developed to remove idiosyncrasy or ambiguity from any report. The Society's rules stated that "in all reports of experiments . . . the matter of fact shall be barely stated, without any preface, apologies, and rhetorical flourishes."

Now that the Society had standardized the reporting of experimental evidence in official phraseology (known, of course, only to members), multiple witnessing could also be conducted by correspondence, using standard descriptions and accompanied by extremely detailed, standard illustration. With these detailed control regulations, the modern scientific paper was born.

Boyle's aim was to create an exclusive, experimental community, with its own forms, conventions, and social relations that would act as a reliable base for encouraging socially acceptable forms of invention and discovery. This activity would demand a new kind of regulated, official workplace where innovation could be carried out according to approved processes and with approved equipment. Boyle called the workplace a "laboratory" and the investigators "priests of nature," whose experiments would be best performed on Sundays and whose labors would produce those "matters of fact" which would best strengthen society ideologically and economically.

In the new laboratory, new-style, "objective" management of data was to be conducted for the first time with the aid of standard, uniform instruments. These would, as Bacon had said, remedy the weaknesses of the human senses and help avoid the metaphysical arguments that had earlier led to the kind of untrustworthy data, which might promote undesirable political and theological heterodoxy.

These new instruments would themselves extend the range of scientific discovery and serve to make it even more exclusive, because in many cases a phenomenon could only be observed and "witnessed" through the collective use of the instrument. It was this collectivity that Boyle saw as the Society's chief virtue, because in a world where it was dangerous not to conform, the Society provided an environment where disputes could occur and subversive errors be corrected quietly, behind closed doors. The language of science would be the language of conformity. No individualism or dogmatism (as defined by the Anglican Church) would be allowed, so only when the community of experimenters agreed did something become a matter of fact.

That the Royal Society was not the democracy of science it pretended publicly to be can be seen in the fact that non-Anglican free-church scientists who were not members of the experimental group were still denied the provisions of the Act of Tolerance established a few years earlier after the Restoration of the monarchy. This Act had been aimed at lifting the most severe restrictions that had been placed on the activities of free-church members after the end of Cromwell's Commonwealth. The restrictions had barred them from membership in universities, military forces, political parties, and major institutions, of which the Royal Society was now one. It was argued that anybody who resisted experimental philosophy, as practiced by the Royal Society, by definition also resisted the established religion. More echoes of Mesopotamia.

Much was about to be denied people who did not conform, because as new kinds of observational instruments proliferated, a completely new kind of knowledge began to emerge in the form of new "instrumental" phenomena, which could only be observed with the aid of instruments like telescopes and microscopes. Bacon and Descartes had enabled the axemakers to create their own "new worlds," known only to those with the equipment to observe them and qualified to work within them.

The proliferation of instrumental investigation began to generate even more new disciplines. By 1673, the Society had many committees, each dealing with the different materials and observations it had requested from all over the world. These now included matters of pharmacy, agriculture, antiques, chronology, history, mathematics, shipbuilding, travel, mechanics, grammar, chemistry, navigation, architecture, hydraulics, meteorology, statistics, longevity, geography, and monsters.

In spite of this apparently wide-ranging free investigation and serendipitous manufacture of new knowledge, the Society was in fact keeping a firm grip on what could and could not be said and done. The case of the vacuum makes the point. Before its discovery and then its confirmation by observation and experiment and multiple witnessing, the vacuum had not existed. Indeed, to have suggested that it did would have been heretical, since the church accepted Aristotle's view that the vacuum was impossible because the movement of a body was slowed down by the presence of air. So in a vacuum, movement would be instantaneous and this was never seen to be so.

In any case, Aristotle had said that space was created by God (whom he called the "Prime Mover") as a receptacle to be occupied by solid bodies, and if any part of space happened to remain unoccupied, as might have appeared to be the case with the vacuum, the omnipresent God would fill this part of space with light. Space was never empty, so the vacuum did not exist. It was not only nature, but it seemed that the church also abhorred a vacuum.

However, in 1635, during the construction of water gardens in Florence, engineers had found that suction pumps supplying the fountain water would not bring up water from deeper than about thirty feet at a time. Galileo was asked to see if he could reproduce the conditions that had created the problem. In 1638, when details of his experiments were published in Rome, they stimulated the Roman Academy professor of mathematics, Giovanni Berti, to conduct a full-scale test.

On the side of his house he fixed a vertical lead pipe, at the top of which was fused a glass flask, with a screw cap. At the bottom of the pipe, Berti attached a brass tap and set the whole apparatus above a wooden cask. He then filled the pipe and the glass flask at the top with water and closed the screw cap on the flask.

When the brass tap at the bottom was opened, the water rushed out of the flask, ran down the pipe, and began to fill the cask, but when the flow had stopped a thirty-foot-high column of water still remained in the pipe, with an empty space above it. When the screwcap was opened again, air could be heard rushing into the flask at the top and the remaining column of water emptied into the cask at the bottom.

In 1641, the matter was referred to Galileo's successor, Evangelista Torricelli, who hit on the idea of using mercury. It was denser than water and would conveniently reduce the scale of the equipment by fourteen times. In 1644, Torricelli was visited by a French experimenter called Marin Mersenne, who lived in Paris and who spent most of his time putting experimenters all over Europe in contact with each other. Mersenne promptly returned to France and reported the news to all his correspondents.

One of them, the French mathematician Blaise Pascal, carried out secret experiments that led him to conclude that the space left in the top of the pipe above the standing column of mercury was indeed a vacuum, and that the height of the column of any liquid depended on the column's support by the air pressing on the surface of the rest of the liquid in the container at the bottom. On the evidence, air pressure was obviously enough to support the weight of a thirty-foot-high column of water, and this explained the original problem with the suction pump.

In 1658, Boyle upended in a dish a three-foot-long tube, filled with mercury, inside a flask which was attached to a vacuum pump. The top of the mercury column in the tube settled at 29 inches. As the pump drew out the air in the flask, the mercury level in the tube fell. When air was once more let into the flask, the mercury column rose again, and when excess air was pumped in, the liquid rose above the original level. This conclusively proved the existence of air pressure and set the scene for major developments in respiratory medicine, pneumatic chemistry, and the investigation of gases.

But Boyle tipped his hand when it came to an "experimental" description of the phenomenon the pump had created. He avoided the theological pitfall of the vacuum by denying its existence and describing it as a space "almost" totally devoid of air, and he did not "dare" to judge whether the space was "devoid of all corporeal substance" or not. Boyle's hidden agenda had been to neutralize politically dangerous views, because if the vacuum really existed, then space was not, as the church taught, filled everywhere by God's presence. And if there were some place in which God could not be present, then how valid was the authority of his royal representative sitting on England's throne, or indeed that of all ecclesiastical authorities?

The first "useful" application of the vacuum pump was in the construction of instruments to measure air pressure, because it had been noticed for some decades that change in the weather caused water levels in tubes to rise and fall. In 1642, Torricelli designed a device consisting of a long tube almost entirely filled with water on which floated a wooden figure. When he set it up on top of his house in Florence, he was able to observe the onset of fine weather and a rise in atmospheric pressure by the appearance of the floating figure above his roof as the water level rose in response. Meteorological observation was soon standardized and facilitated by the production of a portable, mercury-filled version of this apparatus and known as a barometer.

The vacuum also continued to generate new, more esoteric phenomena. Between 1660 and 1663, experiments conducted on birds, snakes, frogs, fish, mice, and insects showed that animals died in a vacuum, and this observation, together with the fact that Boyle was able to use his air pump to extract gases from blood, drove researchers in the direction of respiration and, above all, to investigate the composition of air. As had been previously observed, air was also clearly necessary to combustion. So did it contain a material that aided burning? The theory developed by Mayow, another member of the Royal Society, was that air included a combustive agent, and since gunpowder was the only material that would combust in a vacuum, it was concluded that a substance similar to gunpowder must exist in air. Mayow called this "nitro-aerial particles," and these remarks set off a century of gas experiments that would lay the ground for the modern science of chemistry.

Two other major areas of research came out of the discovery of the vacuum. Boyle collaborated with the Frenchman Denis Papin in experiments on air pressure that led Papin to develop a machine to create a vacuum by condensing steam in an enclosed space. The work laid the ground for the development of the steam engine and the Industrial Revolution only a century after the new scientific method had emerged.

Perhaps the most esoteric product of vacuum experimentation was to come from the barometer. At the time, the instrument was a piece of specialist technology as popular as cellular phones in the modern world. Every "experimenter" worth his salt had to have one. In 1675, a French astronomer called Jean Picard noticed that the mercury in his barometer glowed when the instrument was jolted and the liquid moved up and down against the glass container. This observation led to investigation of the glow and the eventual discovery of electricity.

After the vacuum, the second of the "new worlds" generated by the scientific method was made possible by optical instruments, the telescope, and, to a much greater extent, the microscope. When Galileo used the new Dutch telescope (Hans Lippershey's battlefield "looker," which had been turned down by his patron, Prince Maurice of Nassau, on the ground that he preferred binoculars), he saw a cosmos that had not been known before. This discovery was another example of the way in which the new instrumental sciences, like those associated with the vacuum, downgraded non-axemaker, "unqualified," direct observation by the naked eye.

The telescope showed Galileo the moons of Jupiter, a lunar surface that was not smooth, spots on the face of the Sun, and many more stars than had been thought to exist. These matters of fact were in themselves heretical enough, since according to Rome (e.g., Aristotle) and despite Copernicus, all heavenly bodies had the Earth as their center of orbit, the Moon was a featureless perfect sphere, and the Sun was without blemish.

But when Galileo observed a transit of Venus, proving in this way that the planet was, as Copernicus had claimed, orbiting the Sun, the telescope unambiguously challenged Catholic orthodoxy and accelerated the rate at which science would take power out of the hands of the religious authorities and create a new generation of secular axemaker whose instruments alone could "see" the new truth, because they generated conditions under which it could be seen. From now on, political power would be sustained by science.

A more unexpected world was to be revealed by the microscope. In terms of its absolute novelty, microscopic life was as big a shock to traditional authority as had been the discovery of America and was to have as profound an effect on contemporary thinking. The thousands of new phenomena revealed by the microscope further strengthened secular intellectual independence from the church and drove investigation in many directions at once. Entirely new scientific disciplines emerged as a result of this revelation of new worlds, in which only axemakers would be qualified to operate.

Galileo's microscope revealed the compound eyes of insects, and in 1625 another Italian, Francesco Stelluti, published an account of the anatomy of bees. In 1628, the English anatomist Harvey published on the movement of the heart and the blood, after he was able to examine crustacea, mollusks, and insects "with the aid of a magnifying glass." Harvey's work was to aid the vacuum experimenters in their discovery of the chemical and gas constituents of the blood, and in 1651 he also generated the new discipline of embryology with his microscopic studies described in *The Generation of Animals.*

In 1660, the Italian professor of medicine and future physician to the pope, Marcello Malpighi, used microscopic data to explain the operation of the lungs, to show how the capillaries linked arteries and veins, and to discover the taste buds on the tongue, as well as the cerebral cortex and the existence of red blood cells.

A few years later, a self-taught Dutch amateur called van Leeuwenhoek sent drawings of his microscope observations to the Royal Society. Unfortunately, early on, the Society did not have good enough microscopes to witness his work as a matter of fact, but Leeuwenhoek brought the microscopic world to more general attention with his four-volume work *The Secrets of Nature,* published in 1695 and detailing his microscopic studies since mid-century.

He completed Malpighi's work by showing that arteries and veins led to and from the heart. He drew sketches of red blood cells, showing that they were circular in humans and mammals but oval in fish and amphibians. He produced illustrations of protozoa teeming in a single drop of rainwater, and in 1683 he had examined scrapings from teeth and found bacteria. He also found that aphids reproduce asexually, discovered rotifer organisms, and examined spermatozoa, the lens of the eye, the structure of bones, and yeast cells.

In botany, in 1682 Nehemiah Grew published *The Anatomy of Leaves, Flowers and Fruits,* which was read to the Royal Society. Grew's observations suggested, on the basis of microscopic examination, that leaves were the organs of plant respiration, laying the groundwork for Priestley's later vacuum-chamber work on the respiratory activity of mint. Grew was also the first to speculate on plant sexuality, providing information that would aid Linnaeus in the next century. All the botanical work came together as a formally organized, new discipline when, in 1686, John Ray collated the studies of Grew and Malpighi in a book called *The Natural History of Plants,* in which he based the classification of over three thousand plants on differences among seed types.

So, in a few decades, the microscope had driven knowledge to differentiate into a large number of new specialist sciences, each isolated from the other and from the non-"experimenter" members of the community. Biology, for instance, was no longer a single subject but had split into embryology and developmental studies in general, comparative anatomy, cytology, histology, microbiology, and entomology. Above all, perhaps, the microscope confirmed Descartes' new scientific method because the disciplines it generated were based on the reductionist study of structure, which could be taken apart and put together as Descartes had said, rather than that of process, which could not.

Between them the twin axemaker gifts of the air pump and the microscope for the first time also linked the crafts of engineering and metallurgy with scientific theory. This in turn generated the new occupation of scientific instrument-maker and the new concept of precision. And with the proliferation of esoteric knowledge into so many new theoretical disciplines, demand surged for systems of measurement and quantification. Initially this need was most clearly seen in astronomy, where the drive to produce better lenses in turn encouraged the production of more precise ways of pointing the instruments.

In 1640, William Gascoigne developed the micrometer, which consisted of one fixed and one moveable wire set in the eyepiece. By aligning the fixed wire with one side of a star, or the Moon, for example, and using a precision screw driver to move the second wire so that it lined up with the object's other side, its angular measurement could be read off a scale. In 1667, the French astronomer Jean Picard discovered that the system could also be used to measure how far away an object was, so the telescope was now also a surveying instrument with which to cut and control the countryside with roads and canals, ports and defense works. For the first time, travelers could have accurately drawn maps that would tell them where they were. Communication became easier and commerce expanded as regular coach lines encouraged movement both of goods and people.

Precision changed life at sea, too, as increasing trans-Atlantic trade made more accurate calculation of longitude a matter of urgency. The English government offered a prize (in modern currency, about two million dollars) for a solution to the problem. Hadley in England and Godfrey in Philadelphia took the first step with a navigational instrument called the sextant, which used a telescopic sight to provide very precise readings of star position. The second step was to be able to tell accurately what time it was in Greenwich, England, the standard meridian since 1675, because if at any time on a voyage the altitude of the Sun or a star together with the Moon could be compared with tables showing their position at the same hour in Greenwich, the difference between the two positions would indicate to a navigator how far east or west of Greenwich he was.

In 1735, John Harrison produced a spring-driven marine chronometer that was finally tested during an experimental voyage to Barbados. It showed the time accurate to within fifteen seconds over a five-month period. This meant that over a five-month voyage, navigation could be accomplished accurate to within one mile.

With the aid of the surveying telescope, the sextant, and the chronometer, charts and maps could be more accurate than before, so exploration and trade were made easier. Then, to deal with the massive increase in cargoes of tea, sugar, and tobacco entering Britain, the slide rule was developed to help Customs officers calculate tax. Throughout the eighteenth century, scientific experimentation generated more and more precision instruments, all of them in some way boosting trade and communications, as anemometers measured wind for weather forecasters, navigation instruments made journeys safer, and pyrometers aided furnace men.

Machine tools, lathes for cutting fine screw threads in brass and iron, which could then be used to turn scale-marking devices with great accuracy, helped to make all forms of measurement in navigation, surveying, and cartography even more precise. These would be used to plan and build the canals, roads, railways, and bridges of the Industrial Revolution, powered by the steam engine, itself generated by the vacuum experiments and built with the aid of precision instruments.

Quantification directly affected the populace at large when its techniques took social management to new levels of cut-and-control. William Petty, a member of the Royal Society, who had studied medicine at Oxford and was a member of the original Wadham group, echoed Descartes with his development of "political arithmetic," when he used only "number, weight, and measure" to compile the first proper statistical analysis of the population and the wealth of England, published in 1662.

In 1671, the Dutchman Jan de Witt provided his government with a statistical "life" table on which state annuities could be standardized and offered to investors, who would lend the money to pay for war against the French, an example soon followed in France and England. In the same year, England also centralized customs tax-collection with the Office of Inspector of Exports and Imports, the first national statistics department in Europe, and from then on the figures collected by that office also figured in political and economic negotiations with other states. As will be seen, statistics would soon become another effective tool with which to maintain social order.

An early member of the Royal Society, Samuel Pepys, secretary of the navy, was keen to adopt the new "scientific" approach to administration and set up the first official, standardized classification of ships. Another Royal Society member, John Collins, brought accountancy to the aid of the authorities by suggesting that national wealth could be computed on a giant balance sheet that would form the basis for sound, "scientifically" based political and economic decisions. The opportunities for social management by the new axemaker gift of quantification were too good to miss.

In 1692, Dudley North published a *Discourse on Trade,* directly attributing his method to Descartes, in which he analyzed the relationship between money supply and trade for the first time, and like all "experimenters" he looked at economic activity in terms of "mechanisms." This would eventually lead to successful attempts to reduce all aspects of human behavior to mechanistic processes similar to those of the "hard" sciences.

Quantification went hand in hand with the classification Bacon had urged. As knowledge proliferated (and with it the need for control over its dissemination), science sought precision in the descriptions of the new phenomena being discovered or created and for the new areas of the planet that explorers and traders were discovering. A leading member of the Royal Society, John Wiliness, proposed nothing less than a specialized "philosophical" language, because as Bacon had said, ordinary language was imprecise and, in any case, there were no words with which to describe many of the new things being discovered.

A structure on which such a new language might be based was already available in Aristotle's "Great Chain of Being." In its categorization of every living organism from the simplest slime to humans, the Chain provided a ready-made template to use for the further cutting up of nature. The purpose of classification was that the new taxonomy would provide the base for a new language of science by reducing nature to the smallest (possibly common) elements and identifying them. In doing so, the plan was that taxonomists would reveal the full majesty and detail of God's original and orderly design and underpin social stability at the most fundamental level of all.

In 1668, John Ray drew up a table that culminated in a description of all known plants and suggested an idea that promised greater control over the physical world: that nature was only orderly when it was made so by humans. In the apotheosis of this taxonomic effort, the Swedish botanist Linnaeus' great *Philosophica Botanica*, published in 1751, systematized nature and forced it to obey the rules of logic. Thanks to the new use of numbers, scientific method had reduced nature to something that was now, like the rest of the universe, merely a number of elements behaving according to man-made laws and to be ordered or manipulated at will by humans.

In a final, more general manifestation of its power, the scientific method also generated mechanistic attitudes in the political thinking of seventeenth- and eighteenth-century Europe. Knowledge of the universal law of acceleration, for instance, led people to expect that the progress of society would also accelerate with the passage of time. Regularity and uniformity became the hallmark of a "modern" society. In England, even the financial position of monarch was regularized and made uniform with a royal salary, and national financial affairs were codified and monitored by the first national bank.

Perhaps the most far-reaching social effect of the new reductionist philosophy came in 1776 with the economist Adam Smith's theory of the division of labor. In *Wealth of Nations* Smith expounded a new scientific law of economics, in which market forces regulated the economic activity of a country in a manner reminiscent of Newton's universal law of gravity. By showing the interaction of price with profit, economic growth with wages and employment, and supply with demand, as well as linking consumption and property with the circulation of capital, Smith demonstrated the various parts of a mechanism operating unaffected by bias or party and under the control of an "Invisible Hand." This was, he felt, a force that would always seek equilibrium and could be used for predictable social effect, like the other new laws of nature.

Finally, as the culmination of this ability to create "new worlds," scientific method provided the means to apply mechanistic laws to the operation of entire societies with the political tools provided by John Locke, who used the language of Boyle's laboratory in his writings on social processes. There was, he said, a natural law at work governing the affairs of men, just as it governed the trajectory of a cannonball or the pressure of a gas, and this social law manifested itself in the force of self-interest which ruled the behavior of every individual.

The sole aim of government should be to make sure that nothing constrained this natural force of self-interest. Since its most common expression was ownership of possessions, then the prime responsibility of the state should be to protect individual property, leaving citizens free to concentrate on increasing their wealth. As the social expression of self-interest was best achieved through consensus (an echo of multiple witnessing), the natural tendency of those who owned possessions would be to co-exist by mutual agreement, so the pursuit of self-interest would therefore be communally enlightened, aimed at maximizing both the individual and the common good, through the application of "useful" science for common advancement.

Locke's ideas, born of the scientific method triggered by print and the discovery of America, would find their most powerful expression in America itself, at the birth of the United States, a "modern" nation whose constitution would be framed in the language of the laboratory as a "rational, free society instituted on the basis of natural law, consensus, and self-evident truths." America would become the most powerful nation in history, once the Industrial Revolution had provided the richest society on Earth with new tools for cut-and-control on a planetary scale.

Chapter 7

ROOT AND BRANCH

In 1760 a tulip bulb called "Georgie" was stolen from Samuel Sicklemore, in Ipswich, England. It had been the result of so much painstaking research that the local Society of Florists offered a reward equivalent in modern money to about $150.00 for a single bulb.

Around the same time, Capability Brown, the country's great landscape gardener and self-publicist, was boasting about what he did to the land surrounding a great house, describing it as a literary composition with nature, "setting a comma here, a full-stop there." Brown's contemporary, the poet and satirist Alexander Pope, likened the work more to art: "All gardening is landscape painting—you may distance things by darkening them and by narrowing towards the end in the same manner as they do in painting."

In a single generation since the Scientific Revolution had culminated with Newton, science and technology were already giving us a radically new view of nature by suggesting that it could be "improved." As the full force of the scientific revolution began to take effect, the cutting edge of innovation became sharper and more finely honed than ever before. The new axemaker gifts developed in the Royal Society laboratories were spreading into society, giving governments and institutions the power to change the world with unexpected speed and in unprecedented detail.

By the eighteenth century, technology was able to move from creating artificial phenomena in Royal Society instruments to generating artificial forms of nature in field and garden. It would then move on to harness nature itself so as to provide an entirely new kind of power that would bring radical change to the community, first and most influentially in England, which at this time was more open to change than the rest of Europe, thanks to the constitutional nature of the monarchy and the existence of a strong middle class.

At this time, society everywhere was primarily agricultural and life on the land had altered little since the first Levantine settlements twelve thousand years ago. The early Mediterranean scratch plough had given way to the late-Roman northern-European wheeled version, with a coulter that cut the sod and to some extent turned it over, creating furrows that made heavy soils easier to drain. Some advances also had been made in the use of manure for fertilizer, and since the early Middle Ages the collar had harnessed the pulling power of the horse. But, apart from these minor advances, little had changed.

For centuries, most inhabitants of agriculturally based economies had lived their lives by rote, at nature's command. The typical mid-seventeenth-century English village comprised about ninety families, with 1,300 acres to farm among them. The smallest family land-holding generally consisted of four fields, each made up of a large number of long, narrow strips with no fences between them and some meadowland in common ownership. Total village holdings might vary between one and a hundred acres, but the average was close to fifty-five, made up of eighty-eight strips (not necessarily together) and twelve acres of meadow. Each strip sustained only one crop at a time, and every three years all the land would lie fallow and uncropped so as to recover its fertility.

Nature set the calendar of social activities: in September, ploughing; in October, sowing grain, harrowing, and walking the livestock to market; during the winter months, killing and salting the pig, threshing the grain, and repairing ditches and equipment; in March, sowing beans and barley; in April, grazing animals on common land; in May, management of lambing and calving; in June, ploughing fallow land; in July, mowing the hay; in August, harvesting grain and setting sheep on the stubble. In September, the sequence would begin again. This ancient cycle and the lifestyle of the large majority of the population who lived on the land were both to be totally changed by a new axemaker gift.

It all began with Sir Richard Weston, a land owner from Guildford, Surrey, who returned to England from a visit to Holland in the 1640s and wrote a book about the new specialist crop-rotation techniques he had seen there and which were just what English landowners wanted. The Dutch innovations were new fodder grasses (sanfoin, clover, trefoil, and lucerne) that would feed livestock through the year. Planted on fallow land, they put nitrogen back in the soil (though nobody knew it at the time) and, of all of them, clover was especially good at generating full harvests (which is why "to be in clover" came into the language). Another key arrival from Holland was the turnip, which served to keep weeds under control, grew on bad soil, and fed animals in the winter.

The most popular of the new rotation techniques became known as the "Norfolk" system after the name of the English county where it was first used. Four crops would be sown on the same land in annual succession: wheat, turnips, barley, and clover. The effect of rotation and the new crops was to reduce the amount of land under fallow and to increase animal fodder supply, which meant that more livestock could be raised and in turn provide extra manure. In consequence, yields increased, as did profits, and the system spread rapidly from 1650, to be in general use a hundred years later. It would eventually trigger massive population growth and then feed the increased population, most of whom would by that time be long removed from the land, no longer producers of the food they needed to survive.

The new agricultural techniques made it possible to cultivate previously infertile or uneconomic land, which could now be made profitable enough to be cut up, cleared, and then fenced ("enclosed") for use. The total area of new enclosures increased from 40 percent of the land in England in 1500 to 75 percent in 1700. Enclosure was a more efficient way to use the land than the old-fashioned open field because it allowed more rational consolidation of property. The crazy mix of inheritance laws meant that many farmers owned strip fields scattered across the countryside, so new-style farmers bought up strips from different owners, added newly enclosed areas, and assembled large, productive unified properties.

These techniques were to have profound social effects because enclosure cut off the small cottager from his acres and the sharecropper from his common grazing rights. To some landowners, enclosure also offered valuable control of mineral rights, quarries, or soon-to-be-valuable building land on the edge of towns. Land holdings changed as small landowners sold out to the new, upwardly mobile mercantile class, whose fortunes were based on increased trade because of the advances in shipping and commercial technology described in the previous chapter.

The *nouveaux riches* went into the newly profitable agribusiness for two reasons: to make more money and to improve their social standing. These parvenus changed the character of rural life as, all over the country, farmers who had once been owner-occupiers now became tenant farmers working for wages for a landlord who lived and worked in London.

As a result of these "improvements," the average size of farms grew during the eighteenth century, and the landscape began to look much as it does today. Large farms were a better financial risk: they were more likely to turn a profit from new techniques and improved yields, their owners had greater bargaining power, and they made more efficient use of labor and capital. Above all, enclosed land was more efficient land because it lay fallow for shorter periods, and the stock grazed there was healthier because it was isolated from potentially diseased animals. So in the early eighteenth century the "scientific" experimental techniques for improvement, started by the Royal Society and other organizations, began to make major differences in animal husbandry, and with selective breeding knowledge began to change the shape of animals for the first time perhaps since domestication 12,000 years earlier.

The most famous English breeder was Robert Bakewell of Leicestershire who, in 1745, bred the new Leicester sheep and the new longhorn cow. Another experimenter, Thomas Coke at Holkham, in East Anglia, crossbred Devon cattle and produced shorthorn sheep. He also forced new rams to enormous size on rye grass and clover and then fattened them to the limit on turnips, put them to ewe, and inbred them with their own offspring. The result was an animal that grew very quickly and had a high proportion of meat to fat, so it sold for good money and represented a profitably fast stock turnover because the new breed peaked in two seasons instead of four.

In consequence of these techniques, through the century the average weight of animals arriving at market went up from 28 pounds to 80 pounds. Major landowners like Coke started annual sheep shearing and agricultural shows, where people could display new breeds, or exchange ideas for new techniques and crops. The new, large-scale specialist agribusiness was fatal to smallholders, most of whom became laborers or joined the ranks of the unemployed poor. Without common land to use for grazing, they could no longer feed the cow or chickens whose milk and eggs had kept them alive. Worst hit of all were seasonal laborers, who had previously lived by trapping rabbits, burning charcoal, petty theft, or poaching and whose source of food was enclosed and patrolled by gamekeepers.

The social commentator William Cobbett, riding around England early in the nineteenth century, described the extent of the change. He reported that "a farmer in north of Hampshire, who has nearly 8,000 acres, grows 1,400 acres of wheat and 2,000 acres of barley. He occupies what was formerly 40 farms. Is it any wonder that paupers increase?"

Welfare legislation concerning the poor and destitute, which had always been harsh, now became even more so, as the separation of landed and landless became greater. Justices of the Peace, who were in most cases the local landowner, had summary powers of arrest, whipping, imprisonment, and transportation to the colonies. Punishment for poaching was particularly hard. An alleged poacher could be hanged if he was caught with his face blacked in preparation for a night raid. In 1689, there had been fifty capital offenses, but by 1800 there would be two hundred.

Cobbett put the problem down to the change in rural social relationships and the passing of "resident native gentry, attached to the soil, known to every farmer and laborer from their childhood, frequently mixing with them in those pursuits where all artificial distinctions are lost," and the arrival of new landowners "only now and then residing at all, having no relish for country delights, foreign in their manners, distant and haughty in their behaviour, looking to the soil only for its rents, viewing it as a mere object of speculation, unacquainted with its cultivators . . . relying for influence, not upon the goodwill of the vicinage, but upon the dread of their power."

In this new, structured landscape of great houses, formal gardens, and absentee gentlemen farmer-breeders, the right of movement of the laborer was rigidly controlled. People were only allowed to live in a parish (and in not more than one) by right of birth, marriage, patriality, or if they were employed there as a servant, apprentice, or renter. Certificates of movement were needed for anybody who wanted to move elsewhere, and these were rarely granted. Paupers in receipt of welfare were obliged to wear the letter "P."

In 1801, the writer Arthur Young, who had earlier been a supporter of enclosures, changed his mind when he saw the effect of the new legislation on the ordinary laborer, who now spent most of his spare time drinking: "Go to an alehouse of an old enclosed country, and there you will see the origin of poverty and poor-rates. For whom are they to be sober? For whom are they to save? (Such are their questions.) For the parish? If I am diligent, shall I have leave to build a cottage? If I am sober, shall I have land for a cow? If I am frugal, shall I have half an acre for potatoes? You offer no motives: you have nothing but a parish officer and a work-house. Bring me another pot!"

Meanwhile for the new owners agriculture was becoming more and more profitable, as wheat output increased by 75 percent, barley by 68 percent, and oats by 65 percent, livestock size rose by a quarter, and (in the years since 1500) average crop yield doubled. Writing in 1760, the English novelist Tobias Smollett extolled the virtues of all this profitable control and order: "See the country of England smiling with cultivation: the grounds exhibiting all the perfection of agriculture, parceled out into beautiful enclosures, corn fields, woodland, and commons."

Innovations like black tulips, racehorses, spaniels, and hunting dogs, as well as managed farmland and fertilized crops, nitrogen-fixing fodder crops, and cross-bred farm animals were all examples of the kind of changes that had sprung from the new view of the universe that the previous century's axemakers had generated: that nature obeyed laws that could be used to manipulate nature.

Just as the Swedish botanist Linnaeus had imposed order on nature by giving everything a name, so Capability Brown imposed the same order on landscapes with his "ha-ha's" (the name mimicked the surprise people expressed when they came across them), water-filled ditches, hidden at the end of the great house lawn that kept the animals off the lawn but let them come close enough to represent "wild nature" beyond the borders of a new designer-world.

Lord Shaftesbury was one of the first to put into words the pride in control over nature: "One who aspires to the character of a man of breeding and politeness is careful to form his judgment of arts and sciences upon right models of perfection. . . . Who would not endeavour to force nature in this respect as well?" Alexander Pope described the new notion of playing games with the landscape:

> He gains all ends who pleasingly confounds
> Surprises, varies and conceals the bounds.

According to the master of landscape artistry, William Kent, design was all about disciplined freedom, a vision of the countryside in which nature was "perfected" and her "purest truth" revealed (as was supposed to be the case with Linnaeus' listings of species). The aim was to select representative forms and avoid (or breed out) "accidents" thrown up by unplanned natural processes. In an echo of artificial worlds generated by the scientific instruments of the previous century, Capability Brown arranged the landscape so that a park would be surrounded by an encircling belt of woodland to conceal the disordered reality of the agricultural world beyond it. There might be breaks in his wood, but only if they revealed "pleasing vistas."

These new "landscaping" techniques sometimes required the removal of peasants' hovels if they happened to spoil the view. Viscount Cobham razed the entire Oxfordshire village of Stowe to make way for his park and moved the inhabitants to another village two miles away. Even the woodland itself was to be artificially improved, with trees with lower branches grown to conceal the fence around the park. Within a wood, carefully arranged pathways would conduct the walker to places where set-piece, specially landscaped panoramas would strike the eye. Whenever possible in a view, the middle distance was supposed to include a body of water created by damming or rerouting streams, so as to provide half-hidden watercourses and surprising views of bridges.

Rationalism and order had triumphed, and undisciplined nature was now to be reshaped in a better way. William Kent even invented a new "scientific" law: "Nature abhors straight lines," epitomized in the design of the Serpentine Lake, in London's Hyde Park.

Cut-and-control of nature also brought subdivision and mechanization of the agricultural processes as new machines began to appear —wheeled, mouldboard ploughs, wrought-iron swing ploughs, cast-iron ploughshares. Jethro Tull's horse-drawn seed drill sowed three rows at a time, and there were also horse-driven threshers and automatic winnowing machines. At the end of the eighteenth century, the first automatic reaper appeared with sickles on its wheels.

The new eighteenth-century catchword was "progress." The idea was that deliberate innovation should now be pursued so that everything could be "improved" by the application of rational thought and mechanical principles. In 1753, William Shipley of Northampton proposed the establishment of the first new, official scientific society since the Royal Society had been founded ninety years before. His plan was for a Royal Society for the Encouragement of Arts, Manufacture, and Commerce that would offer prizes for ideas, inventions, and products that would "embolden enterprise, enlarge science and refine art," and by doing so "improve our manufactures and extend our commerce" (in other words, make money).

The Society held its first meeting in 1754, and by 1762 it had 2,500 members, including the painter Joshua Reynolds, the lexicographer Samuel Johnson, and the furniture maker Thomas Chippendale. Early on, the Society concentrated on improvements in agriculture, for which the first prize winner was the Duke of Beaufort, who received a gold medal in 1758 in recognition of having sown twenty-three acres of acorns in Gloucestershire. In 1761, the Society held its first exhibition, and Shipley himself was awarded a silver medal for his "floating light" device, designed to save the lives of sailors washed overboard.

The increasing number of tools becoming available to agriculture made possible much more effective implementation of the concept of "improvement." Reorganizing nature wasn't a new idea. It had first made its mark on the land when medieval axemaker technology had provided Benedictine monks with the means to put into practice their belief that humankind was placed on Earth to carve out a new Paradise, because God had sanctioned human domination and control of nature. In the eighteenth century, the flood of new tools and techniques powerfully reinvigorated this view among churchmen. An early tract by an English Puritan, Ralph Austen, bizarrely entitled "The Spiritual Use of an Orchard," stated: "Fruit trees and other creatures do truly preach the attributes and perfection's of God to us."

Puritans took things a little further. Their view was that just as the wilder parts of nature could be tamed, humans could also be improved, especially if personal desires and tendencies could be controlled. Work improved the character and unremitting toil was humanity's lot and the key to salvation, so persistent application to the new practical acts was conveniently enough regarded as proof of an obedient spirit. Meanwhile, each step in the conquest of nature was a step toward the Millennium and the Second Coming, which would be brought about on a cut-and-controlled Earth through a general improvement of the human condition. Exploitation of the environment was a good thing because nature had been created by God for just such a purpose.

Work was virtuous because it stimulated more work, so untiring activity, productive work, and the rationalization of life according to Puritan ethics (through which social control was made more effective) brought its own reward. God also approved of profit making, and Puritans were taught that it was a parent's duty to bring up a child in some "profitable and lawful calling."

There was no conflict between Protestant belief and capitalism, so the Protestant virtues of diligence, moderation, sobriety, and thrift chimed with the qualities that also led to commercial success. It was foolish, said the English essayist Richard Steele, in *The Tradesman's Calling,* for a man to refuse "to take the advantage which the Providence of God puts into his hands." Success in business could come to be seen as a sign of spiritual grace, granted because a man (or, exceptionally, a woman) had labored hard in a God-given vocation.

These new beliefs harmonized with the changes being brought about by the industrial and economic systems of the time. As scientific believers carved out new man-made worlds of animal breeding and landscape management, their economic counterparts guided the world in a different way with another new tool: capital. Improvement by great landowners, hard work by Protestant artisans, and the voyages of commercial exploration made possible by cartographers all generated large amounts of spare money, much of which went to the socially desirable purchase of land.

The Puritan ethic had been so successful in generating wealth that there was enough spare money in the system to trigger the financial revolution of the eighteenth century, and with it came another new axemaker gift with which to organize and control society. The financial revolution, like the agricultural changes that had preceded it, would further isolate and separate people and make it possible to use money to manipulate their behavior. Capital was an exciting new kind of tool because its potential for self-increase was apparently unlimited, a fact that accorded perfectly with the new scientific concept of an infinite universe and that now brought science and capitalism together in a new dynamic.

The financial institutions set up at the time were primarily inspired by the work of John Locke, who conveniently reconciled, in a new and profitable way, the concepts of universal laws, dominion over nature, and profit. Locke saw the growth of plants and the movement of the sky as evidence of a designed universe working according to laws. Locke believed the world was evidently constructed according to reason and order. Therefore God's design for nature was matched by his design for Man, and "God intends Man to do something." So by discovering natural law and choosing to act according to it, Man would be following the Divine Plan as the result of reasoned thought rather than from blind faith.

Natural law must therefore have been designed by God specifically for the improvement and preservation of humankind. Anything that put this preservation at risk should be avoided or prevented, while anything that aided preservation needed encouragement. Social laws and government should therefore act, above all, to maintain orderly preservation of the individual's existence. Individuals (no fools, they) would realize this and give these social laws their wholehearted support.

Self-interest would above all ensure that people would obey laws that aimed to preserve private property. The only crime in owning property was not to exploit it for the general good, so once life and possessions were secured in this way the individual was free to pursue a good and rational life (in other words: "use it or lose it"). "The great and chief end, therefore," said Locke, "of Men's uniting into commonwealths and putting themselves under government is the preservation of their property."

One early Royal Society mathematician, William Petty, had begun to apply the new "scientific" approach to national financial preservation. In 1665, he had worked out the "value" of the entire population by calculating English consumption of basic commodities. If the six million people consumed goods to the value of 4½ pence a day, then annual national consumption was worth forty million pounds.

Petty reckoned the value of all forms of capital goods in the nation was about two hundred and fifty million pounds, and that these capital goods generated income at an average rate of 6 percent, which worked out at 1.5 million pounds a year. For Petty, the difference between this and the total consumed-goods value of 40 million pounds (i.e., 38.5 million pounds) could only have been generated by labor. So Petty had calculated the national value of labor. His *Political Arithmetick* never claimed to be precise, but for the first time it showed such things might be calculable and that the data might prove useful to governments in their search for evermore efficient ways of controlling the population efficiently and profitably.

Toward the end of the seventeenth century another administrator, Dudley North, took up where Petty had left off. He saw trade as a public tool as well as a private vocation, and in 1692 he published his *Discourse on Trade.* It was clearly rooted in axemaker thinking. North wrote to his brother: "This method of reasoning has been introduced with the new philosophy" (he meant Descartes' Method). "The old" (he meant medieval scholasticism) "dealt in abstracts more than truths, and was employed about forming hypotheses to fit an abundance of precarious and insensible principles, such as the direct or oblique course of atoms in vacuo . . . [but] knowledge in great measure is become mechanical, which word I need not interpret farther than by noting it here means 'built upon clear and evident truths.' "

Dudley applied scientific reasoning to wealth, that came, he said, from production and especially from manufacturing. Dudley proposed another new universal law, a "law of trade," by which all prices were dictated by supply and demand. His mechanistic analysis suggested new laws of money: little trade needed little money; the supply of money depended on how much precious metal there was and how much was coined; interest was a price; nothing could lower interest rates except an increase in capital.

North was the first to construct an analysis, based on general principles, that made possible a simple, mechanistic theory of economics: leave things alone to seek their own "natural" balance. The idea caught on with the capitalist community: "Trade has its principles as other sciences have," said the Bristol merchant John Cary in 1717.

The gift with which to turn these new theories into world-altering reality appeared first in Amsterdam, where the Dutch East India Company had been set up at the beginning of the seventeenth century to look for commercial opportunities in the Far East. The Dutch political authorities also set up a new Exchange Bank to raise and administer the flow of capital to fund the venture, and the Bank did so well it was soon doing the same work for the state and for this was granted the highly profitable monopoly of exchange.

With the backing of the government, the new Bank offered unprecedented financial security, so funds poured into Holland from abroad, even from the English Parliament, the Danish court, and the Venetian Republic. Thanks to this, the Bank was able to provide merchants with the foreign currency they needed for expenses and settlement of bills. The Bank also paid all depositors' bills by transferring written notes for the amount to be debited from their deposit without actually moving the precious metal. The deposits gave Dutch currency the kind of stability that rapidly made Amsterdam the financial center of Europe.

The capital accumulation that followed found plenty of work to do, thanks to the fact that much of Holland was at or below sea level. So the Dutch state borrowed heavily from the Bank to pay for public improvements, like port improvement, polder drainage, canal systems, and land reclamation. Capital was also used to develop specialized national industries like cloth dyeing and finishing, as well as sugar refining and tobacco processing (in preparation for the re-export of these transatlantic commodities to the rest of Europe).

In 1694, the Bank of England was established to do the same service for English finance, to manage all borrowing and lending so as to provide the state with loans to fight wars or finance exploration. By 1720, the existence of the Bank in turn encouraged more specialization in the new world of finance, and private banks were established to concentrate on commerce, agriculture, and overseas ventures.

These in turn triggered other new financial instruments. Money to fund overseas exploration came from the new land registers set up to evaluate property so that the owners could then borrow money against it through the new mortgage companies. Country banks rapidly proliferated (there would be 400 outside London by 1800) to secure local loans, transfer payments to and from London, and help in government tax collection. Above all, the banks raised money to fund an increasing number of public ventures, such as canals and road improvement.

Some of the money raised by these new entities established the new insurance companies set up to take some of the risk out of overseas and domestic ventures. The insurance companies brought to an end the old, unprofitable way maritime trade had been forced to operate in the days before insurance, when in order to protect themselves against piracy ships had been forced to sail in fleets, obliged to travel at the speed of the slowest ship, or else wait weeks for a naval escort. This meant they all arrived at the same time, when their cargoes created oversupply and drove down prices.

While insurance was as old as the Chinese four thousand years before, the first Chamber of Insurance was set up in Amsterdam in 1598, and the idea spread like fire. Insurance was extremely profitable because premiums often exceeded losses by two-thirds. Dutch insurance was so secure and well regulated that even during the Anglo-Dutch War a large number of English ships insured themselves in Holland. Early on, insurance premiums were arrived at by guesswork and haggling. But as the seventeenth century progressed, mathematicians like Blaise Pascal analyzed risk and gave the new entrepreneurs the kind of probability mathematics they could use to calculate the chance of success and the expectation of profit. In order to attract even more backers into commerce, insurance then spread from shipping into fire and life.

Backers were further protected by the emergence of the new limited-responsibility joint-stock company, invented primarily to fund the construction of canals all over England. The idea of the canals was to ship coal more easily because it was too bulky to be moved economically by road, so the new canals were an instant success, and by the end of the eighteenth century, Europe was in the grip of canal mania. Given the immense amounts of capital involved, massive speculation was rife and fortunes were made and lost.

The success of all these new systems for raising and managing capital triggered the emergence of yet another new world comparable with those brought into existence by the scientific instruments of the Royal Society. As early as 1709, Henry St. John referred to "a new interest . . . a sort of property not known twenty years ago." He was referring to the new world of stocks and shares. After early opposition to the transfer of ownership of anything but real property, for the first time people began to make money merely by purchasing shares in commercial ventures with which they had no connection. Thanks to the protection of insurance, lack of expertise no longer constituted a risk.

The trick was to understand stock market activity and to generate wealth just by moving money around. Two more new specialists dominated this scene: the stock-jobber and the insurance underwriter, each of them using the new mechanistic laws of investment and the new probability math to assess and predict the value of what was happening in the real world of trade and commerce.

In this new world, the financial equivalent of the Royal Society's "laboratory" was Jonathan's Coffeehouse, situated in Exchange Alley, London, until 1773 when, after a failed attempt by Parliament to close the market, the brokers moved from Jonathan's to a new coffee house at the corner of Threadneedle St., near where the Bank of England stands today. At first the new house was going to be called "New Jonathan's," but the brokers decided to call it the "Stock Exchange Coffee House" instead. Daily admission was obtainable at a cost of sixpence, and potential investors were attracted by lists of stocks stuck up in the windows of the brokers' offices.

Capital was by now also generating major changes in social behavior as wages altered the nature of work and altered the relationship between worker and employer. Time and effort were increasingly measured not in the old sense, as elements in the expression of mutual responsibilities between employer and employee, but in terms of cash.

As the system matured, it took the usual cut-and-control path. Manipulation of capital and resources fragmented the production process, subdividing each job in scope and level of expertise, deskilling the workers, reducing them to units of production that could be more efficiently used, more easily organized, and less likely to object to or demand changes as long as each man's knowledge was limited to the immediate task at hand. A new kind of life was created: mindless repetition of meaningless tasks set to the speed of a machine.

By mid-eighteenth century, a visitor to a metal works wrote:

> Instead of employing the same hand to finish a button or any other thing, they subdivide into as many different hands as possible, finding beyond doubt that the human faculties by being confined to a repetition of the same thing, become more expeditious and more to be depended on than obliged or suffered to pass from one to another. Thus a button passes through fifty hands, each hand perhaps passes a thousand in a day; likewise, by this means, the work become so simple that five times in six, children of six or eight years old do it as well as men.

This new way of doing things was canonized in 1776 in Adam Smith's *Wealth of Nations,* the factory-owner's handbook, in which Smith used the idea of natural laws to develop a systematic analysis of economics and the creation of wealth. Essential, he thought, was the cutting up of labor, since this represented the true wealth of a nation, essential to the improvement of national productivity. The division of labor was already well entrenched in the larger industries, particularly in ironworks, where the physical separation of water-power, furnaces, forges, and mills had long demanded specialist craftsmen working in different places.

Smith's example of the kind of division he advocated was taken from the pin-making industry:

> I have seen a small manufactory of this kind where ten men only were employed, and where some of them consequently performed two or three distinct operations. But though they were very poor, and therefore but indifferently accommodated with the necessary machinery, they could, when they exerted themselves, make among them about twelve pounds of pins in a day. There are in a pound upwards of four thousand pins of a middling size. Those ten persons, therefore, could make among them upwards of 48,000 pins in a day. Each person, therefore, making a tenth part of 48,000 pins, might be considered as making 4,800 pins in a day. But if they had all wrought separately and independently, and without any of them having been educated to this peculiar business, they certainly could not each of them have made twenty, perhaps not one pin in a day; that is, certainly, not the two hundredth and fortieth, perhaps not the four thousand eight hundredth part of what they are at present capable of performing, in consequence of a proper division and combination of their different operations.

Smith's mathematically expressed laws of production copied those of the earlier Scientific Revolution. According to his new "law," to be applied by governments and institutions with increasing effectiveness, the stimulus for the division of labor was the extent of the market, and this in turn depended on the ease of exchange of goods for capital. Continued growth required an ever-widening market in which transportation and financial instruments were essential tools.

The elements involved in this interaction acted like the Newtonian "forces" of the sciences: the "natural" price of a product covered the cost of production at "natural" rates of wages, plus profit and rent; the "market" price was above or below the "natural" price, depending on the relationship between supply and the demand of customers willing to pay whatever the price was. In times of scarcity of product the price rose.

The prime force driving all this was what Smith called the "self-interest" of the marketplace that governed how prices rose and fell, just like gravity affected the equilibrium between moving bodies. When a price was high, the attraction of profit meant that more goods were made. Expanded supply then reduced the market, so prices fell, and this caused producers to lay off labor, or reduce their capital expenditure. Supply then fell, and in consequence demand picked up, and the cycle of events began again. Everything depended on the "natural law" of the self-interest of all the parties involved: capitalist, producer, labor, buyer. The self-interest force would act like all forces, as an "Invisible hand," guiding everything to "advance the interest of society."

Smith could look around him and see his law in action. The agricultural revolution was increasing crop yields and offering employment to laborers, thanks to a sequence of bumper harvests over thirty years of perfect weather at the beginning of the century. So food was cheap and there were plenty of jobs. So people were confident enough to marry younger and have children. So the population began rapidly to increase and to grow younger. Soon there was a massive increase in demand for household articles and food. So profits rose and were either ploughed back into improving equipment, or else siphoned off to the growing money and share markets in the capital cities where, as the stock market boomed, it attracted capital for further investment and improvement.

The economy was ready for take-off. All that was needed was a new means of production that would be less dependent on the limited number of qualified and skilled workers. It should also be more controllable, more reliable, more reductionist and mechanical, more exactly predictable in terms of output, and more capable of using labor efficiently. Above all it should make use of unskilled labor, as there was no longer time or inclination for the old-style apprenticeships and guild training of earlier, slower days. When it came, the new gift did all this and more. It was steam power.

Adam Smith had shown the way and the technologists were close behind, cutting canals for cheaper transportation of coal to factories noisy with newly invented textile machinery that even an untrained child could (and often did) operate, using the gift of steam power. The Industrial Revolution that steam made possible would be the greatest of all axemaker triumphs so far in history, and it would change the entire world. The Revolution began not across a broad front but from small, craft disciplines within the textile industry. All over the country, men and women in cottages had been doing one or another of the many separate processes needed to make finished cloth: carding, combing, fulling, spinning, weaving. Fragmented in this way, production was scattered, slow, and above all difficult to control and manage.

Technology revolutionized the production of cloth one step at a time. A flying shuttle (carrying the thread back and forth across the warp) sped up the weaver's work and triggered the development of spinning machines on which enough thread could be manufactured to keep up with the new water-powered looms and their faster shuttles. In 1769, the advanced spinning machine gave England the monopoly of cotton production because it automated the thread-making process from raw yarn to finished spindle. The spinning jenny, developed in 1764, did the same thing for lighter weft threads, while the "mule" was a hybrid of both and in 1779 it spun thread that was both light and strong.

These inventions changed textile production because they brought the whole process under one roof in a "manufactory" (later shortened to "factory"). From 1780 to 1812, the number of spindles in England rose from 1.7 million to 5 million, and over 100,000 unskilled people worked in the industry. In order to keep up with the spinning machines, steam-power looms, introduced in 1791, numbered almost 250,000 by 1850.

The Industrial Revolution was the first example of the massive changes that would be triggered by the various disciplines of science and technology, developed thanks to the earlier social-control aims of European "societies for the propagation of knowledge." Working to serve the needs of commerce, these different specialist activities would also combine to produce unexpected social results in an entirely new kind of change-making. As industrial and scientific disciplines proliferated, they began to interact and each time the product of the interaction would be more and more unpredictable. Change would soon become the only constant in life.

The first major example of interactive innovation would be the gift that started the Revolution: James Watt's eighteenth-century steam pumping engine. It needed high-precision piston cylinders that would maintain a perfect seal. These were made by the ironmaster John Wilkinson's new cylinder-boring machine, whose cutting head used a new type of crucible steel originally developed by a clockmaker called Benjamin Huntsman, who had been looking for a better spring when he saw high-temperature fusing techniques in a glassworks, which gave him the idea of doing the same thing with steel.

Huntsman steel was so strong it was perfect for the high-tensile spring needed for John Harrison's sea-going chronometer. The new navigational precision made possible by chronometers encouraged the development of more accurate sextants, and in turn created a demand for precision in scale marking. This was satisfied by Jesse Ramdsen's invention of a system using a small screw set at a tangent to a large circular plate on which instruments (whose scales were to be marked) were placed and then scaled with extreme precision.

This kind of accuracy produced the sophisticated land-measuring instruments that made possible the first ordnance survey work and the great early-nineteenth-century measurement of India. Not even Watt could have foreseen that his mine-draining pump would lead to the map-makers' discovery of Mt. Everest.

Wilkinson's precision cylinder-boring tool, tipped with crucible steel, also bored out the lightweight, thin-barreled guns that allowed

Napoleon to introduce mobile artillery to the battlefield, win every military engagement, and change the political face of Europe. As larger and heavier pieces of iron became available, the new cutting steel also made possible machine tools (lathes and dividing engines and micrometers and screw cutters), which in turn facilitated the manufacture of metal parts so uniform they were interchangeable (a concept first used for muskets in the 1790s by the American Eli Whitney and then to manufacture the machine tools themselves).

Interchangeability in machine parts meant that unskilled workers could maintain and repair machines by simply replacing one interchangeable part with another. These new machines deskilled workers and used them interchangeably as if they themselves were machine parts. The first and best contemporary example of this process and its effects was the production of pulley blocks. These were wooden frames holding the pulley wheels carrying the hauling ropes used on British Navy ships. One hundred thousand pulley blocks were needed each year for these ships and for loading operations at the docks. Three sizes of block were needed, and the production of enough blocks for a year's supply normally took five years.

The English engineer Henry Maudslay developed a new system, based on precision-tool manufacture, to make interchangeable pulley-block parts in a production-line process that divided the operation among 43 machines. In a process of social cut-and-control that would become more and more common, Maudsley's machines produced 130,000 blocks in one year and reduced the labor force from 110 skilled to 10 unskilled men.

The Industrial Revolution was also bringing developments in production-related scientific fields like chemistry and setting the pattern for the kind of scientific innovation that would, from then on, increasingly widen the gap between the axemakers and an ignorant public and change its life in unexpected ways.

For instance, at the beginning of the nineteenth century the new gaslight used coal gas which was given off during the coal-coking, which in turn created huge amounts of coal tar. It was during experiments with this material that in 1857 an English chemist called William Perkin discovered the first artificial aniline dye.

Aniline dyes are a particularly good example of the interactive and unforeseen way scientific and technological discovery is triggered. Perkin had, in fact, been looking for an artificial, chemical version of the natural drug quinine, because English imperial administrators in the tropics were dying in large numbers from malaria. The English had no colonies where natural quinine occurred, since it came from the bark of the chinchona tree grown in Spanish South America and Dutch Java. After months of experiment, Perkin failed to make artificial quinine, but the black sludge he finally produced turned out to be the first synthetic aniline dye.

Other new colors soon emerged from coal tar, and in 1876 the German chemist Heinrich Caro discovered methylene blue. A few years later, some of this dye was accidentally spilled into a bacteria culture dish and preferentially stained only the bacteria. This immediately created a new science, bacteriology, with the first use of the dye by Robert Koch in his hunt for the cholera bacillus. Work on distilling fractions of coal tar also led to the discovery of carbolic acid, used first in antisepsis by surgeons like Lister in Edinburgh, who developed methods of spraying the liquid. At the end of the nineteenth century, the German engineer Wilhelm Maybach used the spray concept to design the first modern carburetor.

As innovation triggered innovation, it must have seemed to leaders of the nineteenth-century West that there was no end to what they could do with the help of their axemakers. All over Europe the land was cut up and fertile, the money supply was growing, the market, like the population, was in constant expansion, the steam engines thundered, and (once gas light made night shifts possible) the factory furnace fires lit up the sky in a never-ending fury of production.

The machines worked unceasingly, unerringly, untiringly. The only gift now required from the axemakers was the one Wordsworth had described, one that would turn people into machines.

Chapter 8

CLASS ACT

The rich man in his castle,
The poor man at his gate,
God made them, high and lowly
And each to his estate.

VICTORIAN HYMN

Throughout history, mysterious axemaker knowledge always strengthened social conformity as at the same time it increasingly distanced the change-makers and their institutional masters from the general public whose lives they controlled. The sheer scale and number of new control systems generated by late-eighteenth-century technologists and entrepreneurs widened this gulf and imposed rigid conformity as never before. Such was the rate of industrial innovation that it would force sudden and fundamental changes on a society politically and administratively unready to deal with them. The changes would in turn bring into being new ways to manipulate the proletariat, because thanks to the factories there was a proletariat to manipulate. The new gift would be an ideological tool for control.

In the early nineteenth century, people started to notice how fast things were changing and to question what this rate of change might mean to their lives. In 1828, a magazine published for English factory workers summed up the growing realization on the part of the masses of how powerless axemaker gifts had made them, because they understood so little of the scientific and technological magic that seemed to change their world every day. "We are born ignorant, brought up ignorant, we live ignorant and die ignorant. We are like men groping in thick darkness."

The Industrial Revolution had sucked millions of country folk into newly industrializing towns at a rate too fast for the urban authorities to control them effectively. The effect of a rapidly rising number of factory workers and the new "unemployed," as well as the unspeakable conditions in which such people were obliged to work and live, and above all the unyielding factory regime that gave them no freedom, no education, and no political power, began to express itself toward the end of the eighteenth century in civil disturbance and riots.

Draconian attempts by the civil authorities to suppress and control this social disruption, as well as the simultaneous rise of an autonomous counterculture initially representing the interests of the workers, would formalize the social division generated by industrializing axemaker gifts in terms predominantly of "class." This would happen first in Britain, because it was there that the full impact of industrialization would first be felt.

In 1798, the British Parliament responded to the riots by suspending the writ of habeas corpus and counseling all Christians to support the suspension as being fully consistent with the principles of "that truly excellent religion which exhorts to content and to submission to the higher powers." The Church of England above all was keen to counteract the revolutionary effect of Tom Paine's *Rights of Man* and his subversive attacks on privilege:

> Man did not enter into society to become worse than he was before, nor to have fewer rights than he had before, but to have those rights better secured. His natural rights are the foundation of all his civil rights. . . . When we survey the wretched condition of man under the monarchical and hereditary systems of Government, dragged from his home by one power, or driven by another, and impoverished by taxes more than by enemies, it becomes evident that those systems are bad, and that a general revolution in the principle and construction of Governments is necessary.

One of the propagandizers working on behalf of the government to neutralize this kind of radicalism was the poet and playwright Hannah More, who wrote about the issue of social discipline in tracts that enjoyed astonishing success, two million being circulated in one year. More's solution to increasing disruption was to teach absolute submission to authority and to encourage Christian resignation in the face of want and adversity. One tract, called "Village

Politics," was written, she said "to counteract the pernicious doctrines, which, owing to the French revolution, were then becoming seriously alarming to all friends of religion and government in every part of Europe."

More also penned a series of pieces, cheap and widely read, called "Stories for the Middle Ranks of Society and Tales for the Common People," written to "improve the habits and raise the principles of the mass of the people at a time when their dangers and temptations, moral and political, were multiplied beyond the example of any other period in history."

From the beginning of the nineteenth century, the middle-class-led Evangelical Movement Moral Crusade also preached discipline and self-control to the lower orders. The English social commentator William Cobbett suggested that their real aim was to "teach the people to starve without making a noise and to keep the poor from cutting the throats of the rich." The major writings of the Evangelicals between 1795 and 1829 all echoed the same theme: "To express moral truths . . . and to deduce from them rules of conduct by which the inhabitants of this country, each in his particular station, may be aided in acquiring knowledge and in the performance of their several duties."

Members of the Evangelical movement deliberately infiltrated banking institutions as well as government and many of them served in the armed forces, seeing themselves in the front ranks of the fight for social stability. Evangelicalism, with its language of war, conflict against evil, and its stress on order and discipline, helped to channel dangerous social dissatisfaction into more acceptable patriotic directions.

But to many of the growing number of political radicals on the Left, the Evangelical "moral reforms" were nothing more than propaganda for an authoritarian and repressive industrial system, a means designed by the forces of law, order, and manufacture as a cut-and-control way to create a sober, disciplined, and obedient working class.

The left-wing reaction to the conditions in which workers found themselves was to organize. By 1818, in the cities and towns of the English Midlands and the North, the radicals had penetrated the manufacturing areas with a network of clubs devoted to political discussion and agitation and closely linked to the new workers' press, of which *Black Dwarf* was the most influential. Their main institutional mechanism, the weekly class meeting, was copied from the Methodists.

At the outset, the movement was not particularly working class in character but populist. In 1819, its goal was the good of "all men," rather than only the welfare of those on the factory floor, so initially the leadership reflected a populist alliance of artisans, craftsmen, small masters, and shopkeepers.

But in 1824 another workers' group, the London Co-operative Society, entered the field and things began to change. In the face of the overwhelming power of Victorian capitalism, the Society fought to maintain the working-class values of mutuality and fellowship and to proclaim nothing less than an alternative social order to that of the capitalists. The Society aimed first of all at removing the workers' dependence on factory pay and company stores by setting up collectively supported shops both as a means of providing goods at the cheapest prices and of using any profits for the good of the movement.

And the aim was not only economic. The statement of intent read: "We claim for the workman the rights of a rational and moral agent . . . [as he is] the being whose exertions produce all the wealth of the world; we claim for the rights of a man, and deprecate the philosophy that would make him an article of mere merchandise to be bought and sold, multiplied and diminished, by no other rules than those which serve to decide the manufacture of a hat."

In 1829, the British Association for the Promotion of Co-operative Knowledge enlarged the scope of the Society and its manifesto found a receptive audience in the manufacturing areas, where an indigenous co-operative had already developed with its own network of publications. It has been estimated that by the end of 1831 over 500 cooperative societies were in existence. The movement published a number of journals, held regional conferences, organized many local meetings, and established local centers. One of these was in Birmingham, where in 1828, the Co-operative Society faced the government with the specter of full-blown socialism when its rules explicitly stated that the Society's goal was "community of Property in Lands and Goods."

The establishment struck back at grass-roots level with the earliest and most effective of the government's "reforms," aimed at counteracting socialist tendencies and implemented in the educational system, where they would face the least well-organized opposition. The axemakers had created a problem. Industrialization and cheap food had brought overcrowding and unemployment. For the authorities, the growing delinquency among out-of-work young people showed a clear need for increased supervision and discipline. So new monitorial schools were set up ostensibly to educate but in fact to train children in the factory disciplines of their parents.

The schools were the brainchild of an ex–Colonial Office administrator in India, Andrew Bell, and were effectively run by the pupils themselves, to train them in the enforcement of the discipline that they would later experience in factories. Schools were divided up into classes, each of which was appointed a boy monitor whose responsibilities were: "[the] morals, improvement, good order and cleanliness of the whole class. It is his duty to make a daily, weekly and monthly report of progress, specify the number of lessons performed, boys present, absent." The monitor was also expected to draw up lists of pages read and words learned by the class and, in this way, the process of education was reduced to a kind of production line.

The teaching of spelling and arithmetic by dictation reduced literacy and numeracy to rote, providing a child with the bare minimum needed to perform factory work. In arithmetic class, a "cypherer" read aloud the sums and the class repeated them in a reiterative process much like the factory work of the children's parents. It was not considered socially safe for working-class boys to become involved in mathematical theory itself, so only arithmetic tables were taught. "By this means," wrote Bell, "any boy of eight years old who can barely read writing and enumerate well is, by means of the guide containing the sums and the key thereto, qualified to teach the first four rules of arithmetic . . . with as much accuracy as mathematicians who have kept school for years."

Like the adults in factories, the children worked in rows, leaving their positions only to go to reading posts, where recitation and chanting were the prime means of learning. The texts they memorized carried the messages of conformity and obedience.

> Twenty pence are one *[shilling]*-and-eightpence,
> Wash your face and comb your hair.
> Thirty pence are two-and-sixpence,
> Every day to school repair.

The widely adopted "Lancaster" method for reading used the same techniques for uniformity inducing: the class text was a single book with triple-size print, whose pages were pasted onto cardboard and hung on the wall. Twenty boys would stand around the board (at their "reading post") and when they had repeated the lesson aloud they would go away to practice spelling while another group came to study the cards. In this way, two hundred boys could repeat the same lesson from one card in three hours.

It was of particular importance to the authorities that writing not be taught to young children: the gift of reading might possibly stir up radical thoughts, but without the ability to write, these could not easily be expressed. As Hannah More said: "I allow of no writing for the poor. My object is not to make fanatics but to train up the lower classes in the habits of industry and piety."

The most successful of all educational innovations, the Sunday school, was introduced with the aim of neutralizing potentially disruptive behavior. Sunday schools were started in 1785 by Robert Raikes (a philanthropist whose work had begun with prisoners in the local jail at Gloucester) to clear the Sunday streets of young people with nothing to do at a time when dangerous libertarian ideas were spreading in the wake of the French Revolution.

The schools were open only on Sundays so that they would not interfere with the pupils' weekday factory work. By 1787, the system was handling 250,000 children, and by mid-nineteenth century over two million, all being taught to read and "instructed in the plain duties of the Christian religion with a particular view to their good and industrious behavior in their future character of laborers and servants." In 1846, the Secretary to the Education Committee of the British Privy Council defined the role of schools as "to raise a new race of working people—cheerful, respectful, hard-working, loyal, pacific."

Meanwhile, attempts were being made to set up a left-wing alternative, inspired early on by Robert Owen, a mill manager in New Lanark, Scotland, who in 1816 had remodeled his factory as a community complete with shops, hospital, and school. The New Lanark experiment was described in his influential tract "A New View of Society," in which Owen stressed the importance of infant schools, of play, and the need for children to be healthy and "active, cheerful, and happy."

However, already in Owen's earliest activities, cut-and-control tendencies can be seen behind the libertarian front. New Lanark included an "Institution for the Formation of Character" whose aim was ambitious by any standards and that was based on Owen's belief that human progress would be impossible without first eliminating ignorance. In the Institution, children learned reading, writing, arithmetic, sewing and knitting and, at the end of their day, the rooms were "cleaned, ventilated, and in winter lighted and heated, and in all respects made comfortable" for the evening session for older children and adults.

Owen's "New View" was the starting point of modern Socialism, and his movement reached its high-water mark in the early 1830s, when working-class "Owenites" were setting up co-operative communities, as well as organizations to promote co-operative knowledge and newspapers to spread Owen's ideology. By 1840, the Manchester Owenites had built their own Hall of Science (an enormous Gothic edifice with seats for 3,000) with funds raised by selling one-pound shares to local working men. Within a few months, the Hall was able to offer its members a rich agenda of activities, from evening classes to Sunday schools for both sexes, to dances, picnics, day outings, an occasional Grand Social Festival with dinner and dancing, to a band that also gave regular Sunday morning performances.

By 1842, similar halls were open in more than two dozen Owenite centres. Each building became the focus of the workers' cultural life. In the Salford Social Institution, "a good book of music was formed, singers were drilled, a hymn book published, a form of service arranged [for] the Sunday meeting." But the hymns were not Christian and the sermons were about social topics. This pilfering of sacramental rites increased the antagonism between the Socialists and the clergy when one Owenite, James Morrison, wrote in 1834: "The services make a powerful impression on the public mind, and produce more permanent and decided results than the most eloquent speeches."

The institutionalizing thrust behind Owenism was demonstrated when he advised other socialists: "Make a new ceremonial for yourselves, a rival ceremonial, which shall win the people on your own side, by the very same arts by which kings and generals, bishops and monks, have in all ages secured popularity to themselves. Theirs was the tyrant's ceremony. Let yours be the people's ceremony."

On the other side of the fence, Victorian authorities were also using church ceremony for indoctrination purposes with hymns that now carried powerful propaganda messages. Because of the sheer volume of official hymnody, this would have a powerful and long-term effect on the general public (even up to the late twentieth century). The 1861 edition of the Anglican *Hymns Ancient and Modern* sold 4.5 million copies in its first seven years.

Hymns counseled the value of submission:

Yet when bowed down beneath the load
By heaven allowed this earthly lot

Look up! Thou'lt reach the blest abode.
Wait, meekly wait, and murmur not.

The value of the work ethic was emphasized:

Happy we live when God doth fill
Our hands with work, our hearts with zeal.
The more we serve the more we love.

As the factory system accustomed more people to time keeping and cut and controlled their days into periods of work and rest, hymns encouraged them to punctuality:

Give every flying minute
Something to keep in store.
Work, for the night is coming
When man works no more.

The Evangelicals used military symbolism to extol struggle and victory:

Fight the good fight with all thy might.
Christ is thy strength and Christ thy right.
Lay hold on life and it shall be
Thy joy and crown eternally.

Hymns exhorted the singer to run straight races, endure suffering nobly, play the game, and be decent and fair. Hymns were even aimed at specific sectors of the population: one non-denominational hymnbook of 1868 addressed each hymn to people in different occupations and social classes. Each group was encouraged to accept their social station meekly, while hard work and social quiescence were repeatedly emphasized. Many hymns took an admonitory attitude, warning the believer against involvement in behavior that might disqualify them from entry to Heaven.

As might be expected, by far the greatest number of hymns were written to indoctrinate the young, at Sunday school, where they would make deep and lasting impressions. Strongly didactic, the verses emphasized the prime value of obedience:

We must meek and gentle be,
Little pain and childish trial
Ever bearing patiently.

Children of desperately poor factory workers were taught that riches made salvation more difficult to attain:

In this our low and poor estate,
Thy mercy, Lord, is clearly shown,
For hads't Thou made us rich and great,
How hard might then our hearts have grown!

Rudyard Kipling's "Father in heav'n" included all the major cut-and-control effectors: patriotism, racism, the urge to charity, the value of work, obedience, and a call to arms in defense of the establishment:

Land of our birth we pledge to thee
Our love and toil in years to be,
When we are grown and take our place
As men and women with our race.
Land of our birth, our faith, our pride,
For whose dear sake our fathers died,
O motherland, we pledge to thee
Head, heart and hand through the years to be.

The Victorian authorities also created a myth with which to further entrench the desire for conformity and obedience, instilling in the middle classes fear of the "residuum," a vast, formless, unidentifiable underclass of paupers that was sickly and dangerous. Social Darwinism gave pseudo-scientific support to the idea that the residuum was something unnatural, teaching that only the fittest should be allowed to survive, and that no attempt should be made to aid the survival of the unfit. So the residuum was defined, by a perversion of this most recent of sciences, to be the product of too much sanitation and charity, which made easier the survival of physically and morally diseased children who would further infect the body social.

In 1884, the Cambridge economist Alfred Marshall suggested that labor camps be set up for the residuum outside London, because there was "no more truly beneficent work than to deprive progress of its partial cruelty by helping away those who lie in the route of its chariot wheels." Crowded into slums where decency and a healthful existence were impossible, this vast population of "degenerates" was said to threaten civilization.

Early in the century it began to dawn on the administrators that increasingly sophisticated technology would work unprofitably unless some efforts were made to train more workers in factory skills. A book called *Practical Observations on the Education of the People* became a runaway bestseller, reaching twenty editions in its first year. It advocated clubs, discussion groups, and libraries where the necessary skills could be taught to workers at the end of the working day. As for more practical instruction, it said, math books should give "only an accurate knowledge of the most useful propositions and their application to practical purposes." Public lectures on mechanical philosophy, chemistry, math, astronomy and geology would be "helpful."

In 1824, the new Mechanics' Institutes, opening in various British cities, held classes from eight to nine PM twice a week for six months and offered lectures on geometry, hydrostatics, and application of chemistry, as well as electricity, astronomy, and French. The Institute's agenda was clear. They would meet the growing demand of the working class for education and knowledge of things other than their immediate working needs, but everything would be kept well in the control of industrialists, who provided the finances.

Some Institutes offered a course in political economy, aimed at correcting workers' "erroneous" views on the nature of capitalism.

In 1827, the pro-capitalist Society for the Diffusion of Useful Knowledge was founded, with a prospectus identifying its aim as "the imparting of useful information to all classes of the community, particularly to such as are unable to avail themselves of experienced teachers, or may prefer learning by themselves." These were the beginnings of middle-class patronage of working-class self-help, and though some of the books published cheaply by the society were useful technical manuals, most of its publications provoked contempt among working-class radicals. In 1832, the *Poor Man's Guardian* described the SDUK as "the disgusting society . . . [which] under the mask of a liberal diffusion of really useful information, has spread abroad more canting, lying, and mischievous trash than, perhaps, any other society that ever existed." The Society's weekly *Penny Magazine* was also bitterly criticized for "containing a pack of nonsensical tittle tattle . . . to the poor and ignorant utterly useless."

In 1829, in a reaction to the propagandizing effects of the new official Institutes, Rowland Detrosier, a self-educated cotton spinner, led a movement to establish a breakaway Mechanics' Institute under working-class control. Like other radicals, he saw lack of popular education as part of a broader pattern of exploitation and deprivation stemming from industrialization. He wrote: "Our labouring population are indeed no longer the serfs of the land, but they are the slaves of commerce, and the victims of bad government."

For the mass of people, the whole of their education "presents to them scarcely any thing more edifying than the examples of ignorance and brutality. The master class is interested in cultivating only one of their virtues, their industry, and to develop that no pains are spared, no means left untried that avarice can dictate, or poverty oblige its victims to submit to." Detrosier added: "To govern is assumed to be the peculiar province of the few; to labour and submit, the becoming duty of the many."

The Victorian authorities' most effective ally in undermining the radicals' claim that left-wing programs offered the only hope of improvement in the workers' lot was Samuel Smiles, whose enormously successful book *Self-Help,* published in 1859, was by far the best-known work of its kind. Smiles saw education as a means to genuine freedom and independence and scornfully dismissed the conservative notion that popular education would lead working men to challenge the established order: "Welcome the education which shall make men respect themselves, and aim at higher privileges and greater liberties than they now enjoy."

Smiles attacked the problem of workers who had been degraded and rendered radical by industrial working conditions and unbearable social circumstances by emphasizing moral and intellectual development, claiming that the end product of self-help would be an individual of unsurpassed nobility of mind and character: "We can elevate the condition of labour by allying it to noble thoughts, which confer a grace upon the lowliest as well as the highest rank . . . Even though self-culture may not bring wealth, it will at all events give one the companionship of elevated thoughts." In a boost to the government's attempt to dilute the effect of anti-capitalist propaganda, Smiles argued that what really mattered was not material success but the development of mind and character.

But Smiles failed to sweep under the carpet the matter of class difference, when his last chapter invited all men to aspire to be true gentlemen: "Riches and rank have no necessary connection with genuine gentlemanly qualities. The poor man may be a true gentleman in spirit and in daily life. He may be honest, truthful, upright, polite, temperate, courageous, self-respecting, and self-helping, that is, be a true gentleman. The poor man with a rich spirit is in all ways superior to the rich man with a poor spirit." The cut-and-control tendency showed through: "The great majority of men, in all times, however enlightened, must necessarily remain working men."

But the problem created by lack of technical education would not go away. The 1884 British Report on Technical Instruction came in response to the growing concern about the capacity of British industry to face foreign (and now especially American) competition. The commission recommended the inclusion in the school syllabus of rudimentary drawing as well as more lessons on craft and agriculture. It also drew attention to the need to train more teachers of science. It urged factory owners to set up schools for their workers. Intelligent children of the artisan class should have access (and scholarships) to technical schools.

By the late nineteenth century, every Western nation had accepted the requirement for technical training and had set up the necessary institutions. In most cases, this greatly increased the influence of the state in the life of the individual, although the change was not always welcomed, even by those in power. The old apprenticeship system had successfully indoctrinated conformity, and social concern was expressed that the new approach might be less effective. But the educational system was rapidly becoming inadequate for the demands of technological advance that ran ahead of the authorities' ability to keep up with the rate of innovation. The railroads, for instance, were creating a huge and unsatisfiable demand for engineers. Axemaker rates of change would outstrip educators from then until modern times.

The new gift of a technological education was not, however, going to be any more available to the majority than schooling before it. Vocational teaching was aimed at creating new forms of exclusive professionalism, and in consequence new middle-class occupations began to form societies to represent and protect their interests. In many cases, in Germany and Britain, the societies provided their members with the means for further self-instruction, as was the case with engineers. There had been few engineers before the mid-eighteenth century, but in 1771 the Society for Civil Engineers was formed, and by 1818 there were enough of them to support a proper professional organization, the Institute of Civil Engineers, which became a prototype for many similar institutions in Britain and the U.S.A. These institutions in turn would make a major contribution to the emergence of an elite managerial class.

Meanwhile the general cut-and-control effects of the Industrial Revolution were changing the shape of society on a wider scale. The new industries had disrupted and eventually destroyed traditional rural social structures. In previous times, individuals had lived within a close-knit, extended family, in country communities where social mobility was limited and where work had primarily involved little more than the provision of subsistence.

The new industrial towns cut off the new village immigrants from nature and from any regard for it, also removing from the new city dwellers any sense of native origin, as well as increasing the isolation of the individual and separating and fragmenting family skills. The factory system also introduced cash wages, and these placed a premium on youth and vigor, and in doing so impaired or destroyed the authority of the old. The nature of urban communities changed as the middle classes left the city centers, not to return there until late in the twentieth century.

The industrialists also changed the individual's concept of time. Previously, work had been defined by the nature of the task and seasonal stages of production would set the rhythm, alternating between times of intense labor and idleness, but now, in the new factories, work was only a matter of how many hours were spent on it and how many units of production resulted. A contemporary Methodist preacher remarked: "I have noticed, too, that machinery seems to lead to habits of calculation . . . In some of the more northern counties this habit of calculation has made them keenly shrewd in many conspicuous ways. Their great co-operative societies would never have arisen to such immense and fruitful development but for the calculating induced by the use of machinery."

Time sheets were introduced at work, and so were time keepers, informers, and fines for lateness, and the factory clock was often locked up so that it could not be altered. In 1770, an early advocate of this time-oriented approach to existence, William Temple, had advocated that poor children be sent at the age of four to workhouses where they should receive two hours' education a day and the rest of the time would be employed in manufactures: "There is considerable use in their being, somehow or not, constantly employed at least twelve hours a day, whether they can earn their living or not; for by these means, we hope that the rising generation will be so habituated to constant employment that it would at length prove agreeable and entertaining to them."

By the end of the nineteenth century, the effects of cut-and-control philosophy had shaped the modern world. Working life was now chopped up and set in orderly sequence, dominated by the need to conform to the machine. At the same time, the proliferation of specialist disciplines had generated an increasing number of esoteric forms of industrial and technical knowledge from which the vast majority of the public was excluded. Education by both church and state served principally to ensure social control through the indoctrination of the virtues of obedience and uniformity.

Against this ranged the emerging and equally conformist alternate systems of Socialism and Communism. In 1884, the Social Democratic Foundation stood for social ownership of the means of production and exchange and saw the retention of political power by the working class as the essential means of achieving this end. In the late 1880s, the formation of major unions of the unskilled took place under socialist leadership, strengthening the connection between political activity and industrial organization that had been lacking in the early days. At the same time, the new socialists embarked on a program of propaganda and education of their own.

It was William Morris in particular who spread the vision of a society where labor would be pleasurable and education the right of all. According to Morris, to achieve a true humanity, people should take pleasure in labor, and this was impossible when the division of labor tied each laborer to a single, detailed operation. The product of such work, Morris said, bore only the mark of its manufacturer, and above all was "of necessity utterly unintelligent, and has no sign of humanity on it." Work of this kind turned the worker into "the perfect machine which it is his ultimate duty to become," so leading to the "complete destruction of individuality."

Morris felt social stability was unlikely while these conditions remained: "To condemn a vast population to lie in South Lancashire while art and education are being furthered in decent places, is like feasting within earshot of a patient on a rack." A new society could only be brought into being by the elimination of class divisions in a society based on co-operation instead of competition: "Let us be fellows, working in the harmony of association for the common good, that is, for the greatest happiness and completest development of every human being in the community." The new society was to be one whose "wealth, resources, and means of production, were owned by the community for the benefit of the whole, ultimately leading to communization of the product of industry as well as the means of production establishing complete equality of condition amongst all men."

For Morris, work was the primary source of human activity, enjoyment, and self-development, so in a socialist society the factory should be a primary educational center. He developed this idea in the article "A Factory As It Might Be," describing a community "where we shall work for livelihood and pleasure and not for profit." Marx had considered this combination of learning and labor as key to the education of the future; Morris took up the idea in the context of his concept of a communal, free society of artists and scientists, where factories would be surrounded by gardens, the buildings would be beautiful, and the workers engaged in honorable and honored labor.

In this way, the socialist factory would provide "work light in duration and not oppressive in kind, education in childhood and youth, serious occupation, amusing relaxation, and more rest for the leisure of the workers . . . [and] that beauty of surroundings, and the power of producing beauty which are sure to be claimed by those who have leisure, education, and a serious occupation."

So in the second half of the nineteenth century, axemaker industrialization had generated two parallel "truths"—socialism and capitalism—and these ideological gifts would cut and control the entire world, dividing it between them for nearly a hundred years.

Meanwhile, by the end of the nineteenth century, the insatiable demand of the new industrial economy for raw materials would lead to the establishment of another new Western world, in the colonies. With all the industrial might of the West, the Christian belief that control of nature was a gift from God could now be applied to the entire planet and its less technologically advanced inhabitants. Many saw it as the West's "manifest destiny" to bring its "superior" Christian way of life to the world and in that way ensure the supply and continued functioning of an industrialized, consumer-led social structure.

Of all the colonialist power elites to answer this new call, the group most certain of the good they were doing, as they cut and controlled the world, destroying its traditions and replacing them with Western models and systems, were the Christian missionaries. As early as the fifteenth century they had spearheaded the drive to convert natives in the newly acquired Americas. This tradition of ownership of newly "discovered" lands was rooted in the fifteenth-century papal partitioning of the Americas between Spain and Portugal.

In the eighteenth-century French West Indies, institutionalized apartheid was justified in an official government memorandum: "Separateness is harsh but necessary in a country where there are fifteen slaves for each white person. One cannot establish sufficient distance between the two species. One cannot instill enough respect in the negroes for those they serve . . . The administration must be careful in severely maintaining his distance and respect."

The French Jesuit Labat wrote that dependence on Western goods was a good technique for creating enslavement, because "they will need them so badly that they cannot be without them and thus will offer . . . all their labor, their trade, and their industry."

Europeans deliberately belittled the abilities of the native societies they subdued and colonies were placed under the control of Western managers, who directed but did not train unskilled and unschooled native labor. These managers then built railroads and ports to facilitate the movement of raw materials, heavy machinery, as well as the army and police.

Above all, the country was always organized along Western administrative lines and according to the location of its strategic reserves and materials, ignoring the social and tribal systems that had existed before. In this way, Western governments effectively neutralized any native managerial and mercantile talent there might have been (particularly in the case of India and Malaya, where native trading organizations had been quite sophisticated).

Once we had in this way denied colonial natives the means to express their own organizational talents, these same natives could then be plausibly described as "disorganized, unproductive, and lazy." It could in turn be argued that such people did not possess, nor could ever have possessed, a system of values. As early as the eighteenth century, the native had been declared by East Indian traders as "insensible to ethics; he represents not only the absence of values, but also the negation of values; he is . . . the corrosive element destroying all that comes near him."

The attitude was particularly effective in regard to the British Indians. Early on, in the eighteenth-century occupation of India by the British and French, Western intellectuals had been fascinated by ancient Indian languages, and native Indian education had even been subsidized by the East India Company. But by the early nineteenth century, Indian superiority in textile manufacture was being neutralized by Western technology, and Britain had begun to look for overseas markets that would absorb its now surplus textile production. It was at this point that Europeans began to propagate the myth that the subjugation of millions of Indians had been managed by a handful of Westerners using better organization and superior knowledge. Indians (and the Chinese too) were now seen as having fallen from their previous ancient state of knowledge, thanks to the deleterious effects of climate or as a result of interbreeding with "inferior" groups.

Europeans were convinced that they had earned the right to take on the "white man's burden" and to become "lords of humankind." Mary Kingsley, an English traveler and writer on Africa (and who purported to admire all things African), wrote on her return from West Africa to Britain in 1895: "All I can say is that when I come back from a spell in Africa, the thing that makes me proud of being one of the English is not the manners or customs up here . . . It is the thing embodied in the great railway engine . . . It is the manifestation of the superiority of my race."

By the late nineteenth century, Westerners had gone through a basic shift in their perception of non-Westerners, such as Australian aborigines, the Amazon Yanomamo, or the African !Kung, who were now all lumped together as "savages" or, according to the catchword of the new anthropology, "primitives."

Western science was seen as an effective tool for weakening native adherence to their own beliefs, and missionaries used the railroad and the telegraph to exalt the Christian God as the only true divine power. One colonial administrator wrote in 1853 that Europeans could "upset the whole theology of the Hindoos by predicting an eclipse." Material backwardness was increasingly equated with heathenism, so mission stations began to teach Western agricultural techniques, and mission hospitals spread "superior" Western concepts of cleanliness and hygiene.

By the 1890s, French political writers like Arthur Girault were claiming the European right to appropriate native resources, which belonged to people who were said to lack energy, initiative, and sense of purpose. Failure to colonize and develop their countries was immoral and against the "workings of the natural order." In 1849, the French novelist Victor Hugo wrote: "France resorts to war . . . only to the extent that it is necessary for civilization. What reassures here is that she knows she bears in her hand light and liberty. She knows that for a savage people, to be occupied by France is to begin to be free, for a city of barbarians to be burned by France is to begin to be enlightened."

In 1878, the British Earl of Carnaervon gave a speech, in which he said: "Vast populations like those of India sitting like children in the shadow of doubt and poverty and sorrow, yet looking up to us for guidance and for help. To them it is our part to give wise laws, good government, and a well-ordered finance . . . It is ours to supply them with a system where the humblest may enjoy freedom from oppression . . . Where the light of religion and morality can penetrate into the darkest dwelling places . . . This is the true strength and meaning of Imperialism."

Behind all the rhetoric lay the real agenda, as Cecil Rhodes (of Rhodesia) expressed it: "I would annex the planets if I could." Europeans needed space in which to settle their now surplus labor force generated by the increasing population triggered by the industrial revolution. Settlement and colonialization would solve the problem of European unemployment and provide jobs for the degenerate urban residuum (who, in going to the colonies, would remove themselves and their diseases from decent society).

Cecil Rhodes wrote: "In order to save the forty million inhabitants of the United Kingdom from a bloody civil war, we colonial statesmen must acquire new lands for settling the surplus population, to provide new markets for the goods produced in the factories and mines . . . If you want to avoid a civil war, you must become imperialists."

By 1880, Africa was the last great prize and it was cut and controlled at the Berlin Conference of 1884. As a leading French economist expressed it: "It is neither natural nor just that the civilized people of the West should be indefinitely crowded together and stifled in the restricted spaces that were their first homes . . . And that they should leave perhaps half the world to small groups of ignorant users who are powerless, who are truly retarded children dispersed over boundless territories."

At the Berlin Conference delegates agreed upon the rules of colonial partitioning so as to avoid conflict among themselves. Any Western country claiming a part of Africa had to declare its intention to annex and follow the declaration with occupation before claim to sovereignty could be considered valid. After occupation, all treaties signed by Europeans with African rulers were legitimate title to sovereignty. By 1910, "title by occupation" was almost complete, as African boundaries were redrawn in Europe with no regard for tribal distribution in those cases where the local chiefs themselves did not know the details (which was most often the case).

It was at this time that the map of the modern Third World was decided. Ultimately, in Africa, this would lead to the establishment of forty-eight states, most of them artificial entities, whose creators had no interest in African ethnocultural, geographic, or ecological reality. Some of the states were giant, like Sudan, Algeria, and Nigeria, while some were tiny, like Gambia, Lesotho, and Burundi. Some had extensive coastline, and some were landlocked, like Mali, Chad, and Uganda, and mineral wealth and fertile land were unevenly distributed, with Algeria and Chad each composed of virtual deserts.

Colonial administrators ensured that the most fertile land was made available to European settlers, except for those areas considered too unhealthy for Europeans, in which cases natives were "assisted" in producing the raw materials needed. In some colonies, notably those held by Belgium and Portugal, natives were forced to produce specific cash crops, such as cotton or sisal, according to very strictly enforced production targets. This work in many cases reduced the extent of local subsistence farming so that food had to be imported from Europe. The price of imports and exports was left entirely in the hands of company officials, and mining especially became the exclusive preserve of Western companies.

The double-think inherent in colonialism had been clearly expressed by another imperialist administrator, James McQueen, back in 1821: "If we really wish to do good in Africa we must teach her savage sons that white men are their superiors. By this charm alone we can ensure their obedience. Without they remain obedient, we will never succeed in rendering them industrious or instructing them in any useful branch of knowledge."

In mid-eighteenth century, the axemakers had provided the means to change the shape of a tulip. Only three generations later their gifts were giving the West the means to change the shape of the planet. The cut-and-control approach to industrial production had also removed and separated most members of European society from their previous direct relationship with the land. In an example to be followed by the rest of the developed world, most of the population now lived in large cities, dependent for survival on cash from a factory that was in turn dependent on raw materials from the colonies. Daily life was now scheduled according to the demands of the factory system and shaped, whether in capitalist or socialist communities, by the decomposition of the community into "units of productive labor."

At the behest of industry, science and technology had already generated hundreds of specialist disciplines, each one of which was beginning to interact with others to bring more and more radical and unexpected change to the everyday life of the ever more uninformed majority. Thanks to the axemakers, the political, financial, and industrial institutions were now capable of shaping every minute of the working individual's life. It only remained to shape the individual's body.

Chapter 9

DOCTOR'S ORDER

Formerly, when religion was strong and science weak, men mistook magic for medicine; now, when science is strong and religion weak, men mistake medicine for magic.

Thomas Szasz

Throughout history the axemakers' gifts have captivated us with their promise. With their carved batons, Paleolithic wizards could indicate the most successful time for hunting because of their knowledge of the seasons. With written schedules for irrigation and for granary inventories, Mesopotamian and Egyptian rulers could promise regular food supplies. Star-reading Greek navigators could assure their masters that cargo ships would arrive safely back in port. The medieval church promised salvation from the pains of Hell, and Industrial Revolution factory owners offered a regular wage.

But perhaps the most seductive proposition came from nineteenth-century medicine, when, for the first time, in return for conformity and obedience, the change-makers offered life. In response to the epidemics that would decimate the burgeoning population, which they themselves had helped to generate, axemakers developed medical techniques that would enable doctors to use the reductionist knife on the human body. The ability to reduce people to numbers and graphs would eventually permit specialists to predict the fate of individuals and communities as accurately as their predecessors foretold the output of factories or the motions of planets.

The first Western attempts to apply these cut-and-control techniques to disease were slow and clumsy, as there were no tools with which to reduce and examine the physical condition, and anyway nobody in 1800 knew what disease actually was. Taxonomy had been tried and had failed because the Cartesian method of reducing a phenomenon to lists of characteristics (in the case of disease, symptoms), however valuable as a means of identifying the presence of different conditions, said little if anything about the disease itself.

The therapeutic value of taxonomic descriptive techniques was dubious at best when, for example, a pathological condition listed as "nostalgia" could only be defined as "an irresistible urge to go home."

There were no means of reducing or subdividing symptoms, or of dissecting living bodies in order to investigate sickness, because disease was not considered to be a localized condition. It was seen as an overall state, a general disturbance of the organism, described by the English physician Sydenham: "It is my opinion that the principal reason of our being yet destitute of an accurate history of diseases proceeds from a general supposition that diseases are no more than the confusion and irregular operations of disordered and debilitated nature."

Before axemaker reductionism had separated them, body and mind were supposed to be one, so the personality and emotional state of the sufferer was of prime concern. Under these circumstances, the patient's view of his or her own condition was virtually the only tool for diagnosis. Since the decision whether or not a patient was sick rested with the patient, the condition presented most frequently to doctors was hypochondriasis.

About all the eighteenth-century medical profession could offer was a good bedside manner. Without proper diagnostic tools and effective cures, the doctor was at the mercy of his patient's expectations, as the French satirist and playwright Molière explained: "The trouble with people of consequence is that when they're ill they absolutely insist on being cured."

A study of anatomy might have been both possible and productive because cadavers were plentifully available, but since each patient's body was regarded as a personal, idiosyncratic entity, the discovery of any general anatomical laws was considered unlikely. In any case, research was regarded as irrelevant, even by major figures like Sydenham, who thought that the job of the doctor was: "to cure disease and do naught else."

The first move towards a properly cut-and-control approach came at the end of the eighteenth century in France, where the Revolutionary wars left so many sick and wounded that the case for effective general therapy became an urgent social priority. Considering that reductionism had originally been triggered in part by the fall of Aristotelian cosmology, it is ironic that the next step in moving medicine in the same direction would be taken by a Frenchman who had previously specialized in celestial mechanics.

In 1798, the first of Pierre Simon Laplace's treatises on the subject were published and made him famous all over Europe. His view of what mathematics could do to aid political authorities in their desire to predict and control social behavior was encouraging:

> We may regard the present state of the universe as the effect of its past and the cause of its future. An intellect that at any given moment knew all the forces that animate nature and the mutual positions of the beings that compose it, if this intellect were vast enough to submit its data to analysis, could condense into a single formula the movement of the greatest bodies of the universe and that of the lightest atoms; for such an intellect nothing could be uncertain; and the future, just like the past, would be present before its eyes.

The key phrase for doctors was: "nothing could be uncertain." Laplacian probability mathematics seemed to offer medicine the chance to find certainty in diagnosis because Laplace claimed to be able to use probability mathematics to extrapolate backwards and derive cause from effect. In 1802, the French government gave him the opportunity to apply his theory on a grand scale.

On September 22 of that year, he made a calculation of the number of births that had taken place in the previous three years among approximately two million citizens. It turned out that there had been one birth per year per 28.352845 people, so if the annual birth rate for the country in those years had been 1.5 million (as Laplace thought), this meant that in all probability the population of the country was 42,529,267. Using his new inverse probability math, Laplace announced that there was only a 1:1,161 chance that his population figure would be off by more than half a million.

However inaccurate the survey might have been, Laplace had invented the concept of the statistically meaningful sample, which would turn out to be one of the most valuable of all gifts for use in social control. The possibility that the behavior of large numbers of people could be reduced to mathematical formulae excited the enthusiastic support of one of the French Enlightenment's leading thinkers, the Marquis de Condorcet, a member of the Academy of Sciences who had a profound influence on the development of a new "science of society." Condorcet was a mathematician whose belief in the inevitable progress of reason led him to publish a manifesto in 1793, in which he said: "We pass by imperceptible gradations from the brute to the savage and from the savage to . . . Newton."

For Condorcet, history was a science whose study was immeasurably aided by the mathematics of probability. Like Laplace, he believed that all phenomena were "equally susceptible to being calculated; and all that is necessary to reduce the whole of nature to laws similar to those which Newton discovered with the aid of calculus is to have a sufficient number of observations and a mathematics that is complex enough." History, above all, was a "science to foresee the progress of the human race" that would make it possible "to tame the future." The agents of progress would be scientists, and so the more support the state provided for education to train more scientists, the more likely that statistical probability would "discover" enough of them in the population to ensure progress. This idea has driven Western planning ever since.

Condorcet thought that the manipulation of masses of people by numbers needed two kinds of data collection: general observation of the whole population (through the application of mathematics) and the intensive examination of a limited number of specimens (through the application of medicine). These studies would make possible the "limitless perfection of human faculties and the social order." Use of the calculus of probability would mathematize social phenomena and introduce predictability and natural law into the behavior of communities. Condorcet trusted in conformity: "Since all men who inhabit the same country have more or less the same needs and since they also generally have the same tastes and the same ideas of utility, what has value for one of them generally has it for all." With the accumulation of enough data, then, and the application of the calculus of probability, the state could be run by social mathematics. For both Laplace and Cordorcet, management of this new science of social decision making should rest only with those capable of understanding the mathematics.

The medical profession was eager to get their hands on the new mathematical techniques, since they might reduce uncertainty through the analytical treatment of a number of individual isolated certainties (patients), which could be compared with others, so that any patterns of convergence thrown up could be used to predict. Clinical medicine could begin to treat the patient as one of a series of endlessly reproducible pathological data to be found in all patients suffering from the same condition.

Fortunately, at least for this enterprise, the French Revolutionary wars had created large enough numbers of sick and wounded for the technique to be applied on a large scale. In 1807, at a time when London hospitals housed about three thousand patients, those of Paris held over thirty-seven thousand, who would be a plentiful source of case histories. The driving force behind the implementation of the new technique was a doctor and government minister called Pierre Cabanis, who in 1798 published an influential paper called "On the Degree of Certainty in Medicine." Cabanis had been a member of an axemaker cabal called the Circle of Auteuil, where thinkers like Franklin, Condillac, Diderot, D'Alembert, and others met regularly.

The situation in which post-Revolutionary hospital patients found themselves was to set the mode of behavior between doctor and patient from then until modern times. In the main, the new patients in the French hospital beds were ignorant, illiterate soldiers, accustomed to military discipline and the imposition of uniform behavior. They had little, if any, sense of privacy and were used to being handled roughly by their superior officers. And since they had lived in army barracks, the lack of privacy in large hospital wards was unexceptional.

With these wounded peasants began the modern reverence for the doctor of medicine, who from then on began to ignore his patient. Medicine was now free to move away from therapy and healing (what the patient wanted) to diagnosis and classification of disease (what the doctor wanted). Once again, the gulf between the expert and the uninitiated would widen.

The passive attitude of the new class of patient set the style of training in clinical medicine, now that trainee surgeons and doctors could poke and prod, undress, examine, and expect instant obedience from their subjects. This fortunate situation—for the doctors—contributed significantly to the rapid spread of medical teaching in schools such as the *Ecole de Santé,* an institution of such prestige that Cabanis turned down the post of Ambassador to the United States to become its first director.

In 1802, one of Cabanis' protégés, Philippe Pinel, wrote *Clinical Medicine,* in which the rules for the use of medical statistics were first set out. Pinel encouraged repeated observation of the patient, as well as the regular and uniform recording of findings and comparison of data over time, just as was the case with work in the hard sciences. Pinel followed the dictum of Laplace: "In order to recognize the best treatment in healing an illness, it suffices to try each of them on the same number of patients while keeping all the circumstances perfectly alike. The superiority of the most advantageous treatment will be manifested more and more as the number of cases increases."

The reduction of the individual to a manipulable numerical unit was given further impetus by the discovery of basic physical units, thanks to the work of Pinel's surgeon pupil, Xavier Bichat. His aim was to find an irreducibly small element in the human body. Between 1800 and 1802, he experimented with over 600 cadavers, cutting them up and then pickling, boiling, frying, freezing, and melting them. Bichat did all this because he was attracted by the new nature-philosophy movement that had started in Germany, partly as a result of the early mathematical work of Gottfried Leibniz. His work on the measurement of infinitesimal changes in the rates of acceleration by planetary bodies in orbit had led him to suppose that it might be possible to use the same calculus to measure infinitesimal units of all forms of matter that might form the basic material of all existence. He named these irreducibly small units "monads," and the nature-philosophy movement saw them as the common substrate of all life, the ultimate link between man and nature.

Bichat thought that these units might be observed and measured as a means of establishing mathematical laws for life-forms in the same way Leibniz and Newton had done for cosmology. Bichat's "monads" were human tissue, which he identified and listed according to twenty-one types whose description was couched in the language of the Industrial Revolution. Bichat described human organs as "the little machines in the great machine" and showed that tissue formed the basic structure of these little machines.

Bichat's *General Anatomy* reads like a treatise on factory production techniques: "Analyze with precision the properties of living tissues; show that each physiological phenomenon ultimately derives from these phenomena; that each pathological phenomenon depends on their augmentation; that each therapeutic phenomenon should effect their return to their natural type, from which they have deviated." Bichat had invented pathological anatomy.

With the aid of statistics and pathology, medicine now began to do to disease what classification had done to botany, what logic had done to argument, and what printing had done to languages. Statistics and pathology defined medicine and its goals according to the terms of its own techniques, so what was not investigated by medicine was not disease. Since the study of disease rather than people involved a distancing of the doctor from the patient, more axemaker gifts soon made this possible.

Thermometers were already being used by the eighteenth century, and the standard body temperature of approximately 98.6°F had already been established. At the end of the nineteenth century, a

Viennese doctor with a penchant for music, Leopold Auenbrugger invented percussion, tapping the chest to identify the position of the heart and reveal something of the state of the lungs. In 1816, the French doctor Théophile Laënnec invented "indirect ascultation," using a rigid cylinder of rolled-up card to identify the noises of the chest and then used post-mortem examinations to correlate specific changes in diseased tissue with symptoms he had identified in living patients. Laënnec was able to use ascultation to identify emphysema, edema, or gangrene of the lungs, pneumonia, and tuberculosis, all without the aid of the patient beyond a few obedient movements.

By 1829, Laënnec's cardboard roll had become the stethoscope and the Paris manual of pathology recommended its use, together with percussion, according to a specific set of general instructions for diagnosis: take the history of the patient and the family, the history of the disease present, the patient's state, detail the termination of the condition, and (of course!) divide each category into several subcategories.

Medicine rapidly found ways to reduce data on the human body to many more subcategories. In 1833, chemistry made possible urinalysis with the first test for albumen. In 1841, Becquerel analyzed the amounts of albumen, uric acid, lactic acid water, and inorganic salts in urine, statistically averaging the amount of secretion over a twenty-four-hour period and arriving at a definition of "healthy" and "diseased" states. In 1855, Vierordt used a system of weights and a stylus to present data on the pulse in the form of a graph, and in 1844 the spirometer was used to measure respiratory function.

But the microscope was by far the most effective technology with which to cut and control the patient. In 1841, Gabriel Andral analyzed blood from its visible properties, microscopic features, and chemical composition, averaged out the proportions of "globules, fibrous material, solids, and water in the sick and healthy," and in this way produced a numerical description of blood that could be used in diagnosis.

Endoscopy let doctors look inside the body. In 1851, Helmholtz developed the ophthalmoscope, and in 1855 a singing teacher in London called Garcia invented the laryngoscope. The otoscope came in 1856, and in 1858 mirrors and paraffin lights were added to all endoscopes. The new watchword was: "Not seeing is not believing." But only the doctor was permitted to look. As a result of these new techniques, the doctor and patient were no longer equals because it was now the doctor and not the patient who decided if disease were present. The doctor no longer consulted the patient except at a superficial level and instruments told him things about the patient that the patient did not know or understand.

In 1868, Karl Wunderlich completed the isolation of the individual by reducing all relevant data to a single, standardized chart on the end of the bed. After studying 25,000 patients, Wunderlich wrote an axemaker classic called *On temperature in diseases,* in which he detailed thirty-two conditions and how they affected the patient's body heat. His dictum, "temperature is the hallmark of disease," expressed what he and many others believed to be a "law of sickness." Wunderlich's chart carried figures on temperature, pulse, and respiration, so that at a glance (even if the patient were unconscious or asleep) any doctor or nurse could monitor progress and decide on treatment.

Medicine was becoming a science, like chemistry or physics, as medical technology made disease tangible, visible, and audible while reducing data on thousands of individuals to uniform charts and pictures. Now the medical profession could enhance their already considerable exclusivity by sharing these esoteric data with each other in specialist publications.

The new cut-and-control techniques only began to have an effect on the life of the public at large after the 1831 outbreak of the first cholera epidemic in Britain, which had already killed fifty million people on its way from India in the previous fourteen years. That Britain should have been particularly prone to cholera's horrifying effect was due to the fact that the country was the most highly industrialized country in the world at the time. The 1800 national population of nine million rose to almost thirteen million by 1850, and between 1801 and 1841 London's population doubled to two million. The population was also mobile and the rural poor flooded into industrial cities looking for work, so disease spread quickly.

By the end of the winter of 1831, cholera had killed over 32,000, and the authorities appeared powerless to stop it. They tried bleeding, purging, sweating, castor oil, brandy and opium, poultices, tobacco enemas, and warm sand. The houses of the victims were whitewashed, drenched with vinegar, turpentine, and spirit of camphor, gunpowder was exploded, and barrels of pitch were burned. But it was all to no avail, and by early 1832 there were riots in the streets as the epidemic continued to spread. From the authorities' point of view, the most disturbing aspect of the disease was the way it seemed to single out the poor and destitute, pushing them toward anarchy.

Various explanations were offered for the disaster. According to a widely publicized Anglican bishop, the disease had been sent to "increase the comforts and improve the moral character of the masses," while for another cleric it was God's judgment on a Protestant country for having flirted with popery. Yet another churchman attributed the cause to electors having voted for Jewish and Free-church political candidates instead of those who were members of the Church of England. But the most intriguing idea came from the Reverend Theophilus Toye, vicar of a church near the epicenter of the outbreak in North East England, who announced that cholera had been sent to deter people from marrying the sisters of their deceased wives.

The fear of the disorder that might follow an epidemic prompted the Victorians to turn to the only other means available of tackling the issue: if the problem could not be solved, at least it could be quantified. In 1834, a statistically inclined public servant called Edwin Chadwick had been appointed secretary to the Poor Law Commission. Then in 1836 the government passed an innocuous bill setting up a national register of births, marriages, and deaths for those not in the established church (mainly Free-church members). Chadwick managed to transform this bill into one that would provide the kind of large-scale, statistical information needed to offer the government a synoptic view of the social situation in the country.

Chadwick succeeded in adding one vital clause in the bill, requiring the notification of cause of death. By 1837, the bill had established the General Registry Office, and data began pouring in from 553 districts throughout Britain. On the strength of the new data and that the incidence of typhus had suddenly surged (14,000 cases in London in 1838), Chadwick persuaded the authorities to sanction a small-scale investigation into the possible relationship between sanitary conditions and disease in five underprivileged areas of the capital: Wapping, Highgate, Stepney, Whitechapel, and Bethnal Green.

The report, published later that year, supported the generally held view that the poor drank too much, neglected vaccination, were reluctant to enter a hospital, and didn't wash enough. But the report also suggested that the possible cause of their vulnerability to disease might be the general conditions under which they lived, surrounded by filth in cesspools, privies, and stagnant drains, breathing in the exhalations from undrained marshland, graveyards, and slaughterhouses, and with limited or non-existent access to clean water. The report noted that conditions could be improved by sewer building, provision of water supplies and garbage removal, effective building regulations, and ordinances to prevent overcrowding.

The authorities were alarmed enough to give Chadwick the go-ahead for a more ambitious, nationwide report on the sanitary conditions of the "labouring classes." When this report was published in 1842, it shocked Victorian society to the core and led to measures that would set the style for a radically new relationship between the state and the citizen, first in Britain and then in every Western country. These measures would give public institutions unprecedented powers over the private life of the individual.

What shook the Victorian middle class was that Chadwick's report revealed appalling social conditions in the industrialized cities. The sewage systems were totally inadequate, built as they had been for city populations many times smaller than those now using them and generally consisting of immense, flat-bottomed brick caverns that were washed by a trickle of water and built principally to accumulate deposits. Every five or ten years, sewage workers would break through the brickwork, scoop up the contents, and leave it on the streets for scavenger carts, in what were euphemistically called "golden mounds," some of which rose several stories high before they were carted away to be dumped on fields.

Personal sanitation among the poor was almost non-existent, when typically, three houses would share one of the privies clustered alongside open ditches running along the center of town streets. More than half the privately supplied water was passed untreated and dirty from river to consumer. The majority of the poor had no piped water and were obliged to stand in line at public pumps, which turned on for two hours a day and were supplied with water from wells that were in many cases infected by overflow from cesspits. Even the better off were little better off. In Edinburgh, the water-supply company delivered one gallon a day per person, and most people hoarded water and used it repeatedly until it was only fit for cleaning floors.

Accommodation was equally bad, as in some cities forty people lived in a single house, two families to a room, and it was reckoned that in Liverpool 40,000 people lived twelve to a cellar. In 1840, there were 12,000 head of cattle in London, and in Reading, a major city near the capital, nearly four hundred pigsties were scattered among 2,500 houses. Pigs and cows were kept in the same crowded and filthy tenement courts where children played, and as a consequence tuberculosis, foot and mouth, and lung diseases were rampant.

Street by street and town by town, Chadwick's analysis, accompanied by statistics and illustrations, showed that there was rampant disease, infection, child death, widowhood, and orphanhood, and that the life of the poor was reduced by at least ten years due to lack of sanitation and polluted water supplies. But middle-class society was shocked most of all by the moral problems made explicit in the reports: frequent bastardy, prevalent incest, and children forced onto the streets as beggars and prostitutes. The report suggested that urgent reform was necessary if revolution was to be avoided.

That these reports were able to provide statistical evidence of any kind at all was due, in the main, to the efforts of Chadwick's assistant William Farr, who had studied the new statistical medicine in Paris. Farr standardized vital statistics and provided the authorities with a valuable cut-and-control tool that he called a "biometer." With this, he analyzed data from what he described as a "healthy" district (where the death rate was seventeen per thousand) by categories: age, the number of people reaching that age, the number dead at that age, and rate of death and expectation of life at all ages. The data showed that there was a clear link between mean life-expectancy, numbers living, births, death, and rate of deaths. If a district varied from this pattern, then the situation was due to what Farr described as "preventable" conditions (by which he meant lack of hygiene, overcrowding, inadequate diet, etc.).

Farr used statistics to show, with fair certainty, that the closer people lived to a river, the more they were likely to suffer from cholera. His report came just in time for the next cholera epidemic in 1848. This one was much worse than the first, killing nearly 70,000 people and galvanizing the authorities into passing the Public Health Act and the Nuisances Removal and Diseases Prevention Act. These new laws gave the government new compulsory powers, in an emergency, to effect nuisance removal, street cleansing, house disinfection (with or without consent), or removal of an infected person to an isolation hospital (also with or without consent). Public health considerations were to bring direct state intervention in the private life of the individual.

The fear that cholera outbreaks would occur in "noxious and remote" districts but then spread to wealthier areas generated a new propaganda offensive. Lectures at the new Working Men's College in London now included talks on "God's law and man's law as they affect health and disease," as well as one on "Cholera and social disorder." In 1861, the Ladies' Association for the Diffusion of Sanitary Knowledge distributed 140,000 tracts on "The Power of Soap and Water." In a pastoral letter, the Bishop of London warned that it was "certain that persons immersed in hopeless misery and filth were for the most part inaccessible to the gospel." In 1856, the excise tax on soap had been removed, the price fell by a third, and by 1861 consumption had doubled. Cleanliness was now next to Godliness, and as the German sanitarian Treitsche told his Berlin University class: "The English think soap is civilization."

The health propaganda machine went into high gear and hygiene now became a separate discipline within medicine. The idea was spread around that clean skin was supple and functioned and "breathed" better than dirty skin, giving "an infinitely more restorative rest, which gives the whole body new vigour and energy." A French doctor called Clerget noted that those who had suffered most were those who paid least attention to the rules of hygiene and cleanliness. Clerget was sure cleanliness would gradually bring social order: "Cleanliness calls to cleanliness, clean houses demand clean clothes, clean bodies and, in consequence, clean morals."

An English hygiene tract entitled "Dirt and a Word about Washing" expostulated: "The poor mechanic who takes the pitiful tallow and dirty ashes and changes them into dirt-destroying soap is doing a noble work. It is like what the Divine Being does in nature . . . Dirt is poison . . . It mixes with the blood, it makes it corrupt . . . An habitually dirty man can hardly be religious. He is breaking one of Nature's first laws. Cleanliness in a person prepares for purity of heart."

The other approach to the problem of public health was to try to improve the condition of the individual through regimented activity in the open air, with the hope that in removing themselves from the crowded city environment people would also remove themselves from contagion. In the 1860s, athletics became fashionable, especially in schools, and the British Schools' Boards wrote: "Manly sports, played as they should be played, tend to develop unselfish pluck, determination, self-control, and public spirit." When children played cricket "in becoming better cricketers they become better boys."

At private schools, games became codified and compulsory. The aim, as ever, was to teach the collective qualities of obedience, physical commitment, acceptance of rules, teamwork, endurance, and fortitude on the grounds that "brave boys make brave men." The new sports (and the introduction of drill in classrooms) would also improve the physical condition of men who might be needed for military service. Besides, as Chadwick had argued, if children were fitter "three boys might do the work of five in a factory."

This emphasis on orderly sport was promoted, particularly in Britain and America, by a new movement known as "Muscular Christianity." A number of British private schools were founded by the movement and they introduced the idea, which then became widespread, of devoting half of each school day to outdoor sports. God, apparently, loved an athlete (and besides, too many recruits had been rejected for military service during the Crimean War on the grounds of physical deficiency). As Lord Roseberry said: "It is of no use having an Empire without an Imperial race."

Meanwhile the epidemic of 1853 had triggered further investigation into the root cause of cholera. John Snow's meticulous statistical study a year later showed that water from the fecally contaminated London reaches of the River Thames was nine times more likely to cause cholera than water drawn from an uncontaminated source upstream from the city. Snow also proved conclusively that when water supplies were filtered, death rates dropped dramatically.

In 1866, the final major cholera epidemic killed over 14,000, and in support of Snow's theory it was found that more than a third of the deaths were in those London areas that still lacked filtered water supplies. Although all along, nobody had known what the disease was, attempts at control through quantification by statistics and draconian social control seemed to have worked. The success set the pattern for future state intervention in cases of what was now to be defined as "matters of public concern."

The value of statistics and probability mathematics in making easier the determination, prediction, and control of social behavior had been given special importance by the cholera crisis and the risk of widespread social instability it might have brought in its wake. By the beginning of the nineteenth century, Germany led the field in statistics, and cast its data-collecting net widest to gain information that might be relevant to the more efficient management of social behavior. German authorities discovered that their power to cut and control the community was immensely strengthened by data on geography, climate, economics, agriculture, demography, illness, and natural history. The word "stat-istics" echoes this use of numbers by the state.

In 1826, Moreau de Jonnes wrote that statistics were like "the hieroglyphs of ancient Egypt, where the lessons of history, the precepts of wisdom, and the secrets of the future were concealed in mysterious characters." Part of the nineteenth-century rush to statistics had undoubtedly been due to the social disorder following the French Revolution, while in Britain statistics seemed to offer a tool with which to control the social dislocation caused by overrapid industrialization.

The man to save the day for the authorities was Adolphe Quetelet, a Belgian mathematician who had studied astronomy in Paris in the 1820s and was influenced above all by Laplace. In 1835, Quetelet's *Man and the Development of His Faculties: A Social Physics* became an instant bestseller. In it, Quetelet described the statistical tool that would make social cut-and-control more accurate than ever before, because he found a way to reduce individuality to a norm, which he called "the average man."

Using terminology and analogies reminiscent of astronomy, he attempted to calculate the law governing social behavior in terms of the laws of error originally formulated by astronomers and mathematicians to correct inaccuracies created by imprecise instruments and subjective bias in the observation of celestial phenomena. The average man, Quetelet said, was the social analogue of a physical center of gravity, "the mean, round which the social elements fluctuate." Quetelet was convinced of the predictive power of his system: "We shall be able to fix the laws to which [man] has been subject in different nations since his birth, that is to say we shall be able to follow the course of the centers of gravity of every part of this system."

Quetelet studied statistics on French crime and claimed: "We can tell beforehand how many will be forgers, how many poisoners, almost as one can foretell the numbers of births and deaths . . . Society contains the germs of all the crimes that will be committed, as well as the conditions under which they develop. It is society that prepares the ground for them, and the criminal is the instrument." He produced a statistical law that claimed to predict how behavioral regularities would continue in the future, arguing that these arose from an underlying uniformity in the condition and dynamic of society.

In 1833, the first permanent Committee on Statistics was formed in Cambridge, and Quetelet proposed that the British should establish a national Statistical Society, which was then formed in 1834 and stimulated other countries to follow. The first international statistics conference was to be held in Brussels in 1853.

The London Society showed its particular interest in social control: "statistics seeks only to collect, range and compare that class of facts which alone can form the basis of correct conclusions with respect to social and political government. . . . Like other sciences, [it] seeks to deduce from well-established facts, certain general principles which interest and affect mankind." Quetelet's efforts followed the general attempt to find a way to offer scientifically reliable means of achieving social order through the predictive manipulation of mass behavior. This new science would become known as "sociology."

The epidemics and the social conditions that had made their effect so devastating had both served to strengthen the grip of the state on the community at large. Now the institutionalizing of public health, aided by developments in medical technology, would strengthen it further.

The drive to find the origin of disease itself would also further reduce the status of the individual. The tool to make this possible was produced between 1825 and 1830 by J. J. Lister, a London wine merchant and optical enthusiast, who solved the major contemporary problem in microscopy. Imperfections in lenses created comas, or blurred areas, either in the center or around the edge of the image, and this had led to serious misconceptions by those using the instruments. For instance, in 1823 Milne-Edwards had claimed that all tissue, regardless of type, was made up of globules approximately 1/300th mm in diameter. Lister's improved microscope used a series of lenses whose varied curvature canceled each other's aberrations. In 1827, the first use of the Lister triple-lens compound microscope revealed that blood corpuscles were not globular but biconcave.

Discoveries followed thick and fast. In 1831, Matthias Schleiden saw the cell nucleus clearly for the first time. In 1834, ciliary movement in bacteria was identified by Purkinje, and at the same time Theodor Schwann examined every known tissue and found confirmation of what looked like the basic unit so keenly sought by medical nature-philosophers. In 1839 he said: "There is one universal principle of development for the elementary parts of organisms, however different, and that principle is the formation of cells." The existence of cells showed that life was not a phenomenon involving some unmeasurable "life force," but a quantifiable (and manipulable) entity like those revealed by the physical sciences.

The key advance in cellular pathology was made by another German political radical and physician called Rudolf Virchow, whose influence was so great he was known as the "Pope of German medicine." In 1848 he said: "The physicians are the natural attorneys of the poor and social problems should largely be solved by them . . . Medicine is a social science and politics is nothing but medicine on a larger scale."

In 1845, he delivered a classic axemaker paper on "The need for and correctness of medicine based on a mechanistic approach," which held that life was essentially nothing more than cellular activity. Cells, he held, were the basic unit of existence, and life was no more than the sum of cellular phenomena that could now be submitted to normal physical and chemical laws. And if all life were cells, then disease (being alterations in the cells) was nothing more than life under altered conditions. So if these conditions could be avoided, public health would be dramatically improved. Thanks to Virchow, the microscope would become the new tool for social control.

Virchow made a further link with the physical sciences by describing the cell as "the organic molecule, analogous to the chemical or physical atom . . . We can go no farther than the cell. It is the final and constantly present link in the great chain of mutually subordinated structures comprising the human body." Virchow and his contemporaries thought medicine had finally reached its reductionist goal, but there was more to come.

As pathologists learned about the microscopic aspects of disease, they exercised an increasing control over the hospital clinicians. By the 1860s, the division of labor between the two groups was well established and moving both disciplines toward even greater specialization. As soon as the clinicians realized their diagnoses might be subject to correction by pathological evidence, the more technologically advanced hospitals began to organize "pathological conferences," where clinicians submitted diagnoses to pathologists for confirmation. Diagnosis moved several steps further away from the patient.

The social overtones of Virchow's views of the relationship between medicine and the state were emphasized by his pupil Ernst Haeckel, whose work was to influence Hitler. Haeckel spoke of cells as law-abiding citizens in an orderly *Kulturstaat,* growing and becoming more powerful thanks to the division of labor. According to Haeckel, cells formed "republics in plants and monarchies in animals," organs were like departments of state, and the whole organism was centrally ruled via the nervous system. Through Haeckel, social language entered the medical vocabulary in terms like "cell territories," "cultures," "colonies," and "cell migration."

The final move to reduce the patient to a statistic was taken by Robert Koch, a general practitioner working in a small town in East Prussia, who had an interest in anthrax and who was to bestow medicine's greatest gift on an unsuspecting public. Earlier, in 1850, a French parasitologist called Davaine had reported the transmission of anthrax in sheep via the blood of animals already dying of the disease. Davaine had found microscopic, rod-shaped organisms in the blood of the dead sheep and confirmed that sheep whose blood did not carry these rod-shaped organisms did not become sick.

In 1876, Koch isolated the anthrax bacillus, cultured it, watched the bacillus produce spores in the tissue of animals, and saw the spores in turn produce the bacillus again. Anthrax spores, Koch discovered, could remain in the ground for months before entering an animal's blood and reactivating the bacillus, whereupon the animal developed anthrax. Koch spent three years examining every aspect of the life cycle of the bacillus and showed that he could infect a healthy animal at will in experiments providing the first clear evidence that a specific microorganism caused a specific disease.

In 1878, Koch published on the bacteriology of infected wounds and, realizing that large numbers of bacteria might be involved, put forward a number of rules to aid investigators. Known as "Koch's Postulates," these rules defined all future guidelines for the investigation of disease: a microbe could be accepted as the cause of a disease only if the microbe were present in unusual numbers when the disease was active; if it could be isolated from the diseased patient; and if it caused the disease when inoculated into a healthy subject.

When Koch also developed new techniques for growing bacteria on agar jelly substrates, he gave medicine the ability to manipulate nature at the microscopic level. This did away with any need for the patient's presence, because culturing the disease and then controlling its development required no more than a single drop of blood.

Koch's first major success came in 1882 when he announced the results of his work on tuberculosis. On what is described as a "red-letter day in the history of bacteriology," Koch told a meeting of the Berlin Physiological Society how, in six months' attempts to isolate and identify the bacillus (which was particularly difficult to find and to culture) and working according to his own postulates, he had proved that the bacillus caused the disease. A year later when Koch used the same techniques in India to find the cholera bacillus, he linked the two cultures, microscopic and social, by positively identifying the source of the bacillus in infected stagnant ponds that had been used for drinking water and washing.

By the end of the century, Koch's new science of bacteriology had identified the microbes responsible for tuberculosis, pneumonia, bubonic plague, anthrax, typhoid fever, tetanus, diphtheria, gonorrhea, cholera, influenza, leprosy, actinomycosis, malaria, amebic dysentery, relapsing fever, and trypanosomiasis.

Koch's colleague, Paul Ehrlich, also discovered that some bacteria-staining chemicals would also selectively kill certain bacteria, so he used them to develop the first "magic bullet" drug, Salvarsan, to treat syphilis. Ehrlich also introduced chemotherapy and triggered the modern view that the center of attention for medicine should be the disease and not the person.

Thanks to bacteriological techniques, diagnosis was now able to quit the hospital ward entirely. As the great French diagnostician Claude Bernard had said in 1877: "Well endowed with data collected in hospital, medicine can now leave it for the laboratory." At first the new pathology laboratories were in the hospitals: at St George's, London, at Bellevue, New York, and at Johns Hopkins in Baltimore. But by 1893 increasing costs soon led to the establishment of public laboratories, the first in New York, where diphtheria diagnosis took place. Tubes containing nutrient were distributed free to doctors, who drew a swab taken from an infected throat across the culture and delivered the tube to the nearest apothecary shop. The tube was collected and the next day the doctor received a lab report. In its first three months the new laboratory helped the New York Public Health authorities to identify and treat 301 diphtheria cases out of 431, and by 1895 the lab was also doing tests for tuberculosis.

Public health laboratories brought the combined gifts of microscopes and bacteriology to the control of public health and did away with patient involvement. There was now a growing number of specialists and institutions, concerned with the disease and its behavior, for whom patients provided little more than a source of material for study.

Successes in the labs gave impetus to efforts in Public Health to cut and control disease in the population at large, thanks to the data available from sanitary engineers, physicians, epidemiologists, vital statisticians, lawyers, nurses, and administrators. Americans who had studied in Germany brought back and taught the new bacteriological methods. By 1901, bacteriology was being employed to study water supplies and sewage disposal in Massachusetts. The study would show that typhoid was transmitted in polluted water and as a result would trigger the development of quantitative methods for measuring the presence of bacteria in air, water, and milk.

One key social effect of the labs was to divert the attention of authorities from the larger, more diffuse public problems of water supply, street cleansing, housing reform, and living conditions. Bacteriology was more "modern," and it was also easier and cheaper to administer than the more traditional approach to public health. The new epidemiological view was expressed by the director of epidemiology of the Minnesota Board of Health in an influential book, *The New Public Health.* In it he claimed that modern scientific methods were far better than the old-fashioned, social-reform approach to epidemics. To control tuberculosis it was not necessary to improve the living conditions of a hundred million Americans but merely to supervise and control the 200,000 active cases and limit their potential to infect others.

In 1915, in the first Handbook for U.S. Public Health, almost half the book was devoted to contagious diseases, with much smaller sections on industrial hygiene, housing, water supply, and public education. This narrow, bacteriological view of disease would remain dominant for decades.

So thanks to cholera, nineteenth-century medical axemakers provided a new means of social control through public health measures that virtually removed the individual's right to privacy in matters defined (by the state) as of "public concern." Only religion had previously penetrated so far into the life of the individual. Cut-and-control ideology fitted medicine like a glove, because diagnosis and treatment required conformity from medical staff and obedience from the patient.

Today, the seductive attraction of the physician's world, with its white coats, gleaming instruments, and life-saving gifts has succeeded in "medicalizing" society, as the language and ethics of the hospital have entered daily life. The doctor is the new shaman, closely associated with the materialist values of constantly rising standards of living, better individual health, and the growth of community care. Above all, the doctor represents a "scientific," objective way to judge social behavior (in terms reminiscent of the vocabulary of medicine and statistics) as "healthy" or "diseased" or "abnormal."

Ironically, while medicine is the specialist activity most closely linked to the personal concerns of the individual, it is perhaps more esoteric and exclusive than any other scientific field of endeavor since those most concerned with health (the patients) have become those least involved.

Thanks to the bacteriological revolution and a reductionist concentration on microscopic phenomena, the human dimension of health is today judged as unquantifiable by vast numbers of people and has practically disappeared from medicine. There is minimal place in medical science for the whole human and even less for those non-Western cultures and their non-reductionist understanding of the relationship between individual health and the environment, because Western axemakers have almost completely severed the link between the two.

With Western medicine we have cut apart the human body in the same way as we have, through history, separated the axemaker from non-axemaker, subject from god-king, communicant from priest, nation from nation, and people from land. We have dissected and divided up the world and its inhabitants so that they can be manipulated as economic and political units interchangeable with one another. In doing all this, we have cut out the individual in the same way as we have chopped up the planet, axing the single parts without regard for the whole. As we shall see, this process has brought us close to catastrophe.

III.

PICKING UP THE PIECES

C h a p t e r 1 0

JOURNEY'S END

What now remains of the once rich land is like a skeleton of a sick man, all the fat and soft earth having wasted away, only the bare framework is left. Formerly, many of the present mountains were arable hills, the present marshes were plains full of rich soil; hills were once covered with forests and produced boundless pasturage that now produce only food for bees. Moreover, the land was enriched by yearly rains which were not lost, as now by flowing from the bare land into the sea.

PLATO (FIFTH CENTURY B.C.E.)

For about the last 120,000 years, the axemakers have helped us make a great journey to colonize the Earth and today we have come close to that journey's end. Along the way, on many occasions when the community faced a challenge, the axemakers offered gifts with which we might solve whatever problem had arisen. Our leaders and institutions always accepted the gifts for their short-term value and ignored their long-term cost. In the circumstances, what else would anybody have done? The problem was that one gift led to another. Over time, the axemakers became indispensable and their gifts too attractive to refuse, as events each time conspired to convince us that the name of the game was: "Use it or lose it all."

This first of those many gifts came when the axe introduced a new kind of once-and-for-all change in the cyclical processes of nature. The axe also redirected the development and the selection of individuals in human societies, above all triggering the emergence of institutions like chieftainship and organized religion, whose exclusive use of axemaker knowledge would give them mastery over nature and control over the community. At their command, axes cut a swathe through the environment wherever humans went, clearing our path, making our progress easier.

In the beginning, the marks that our axes made were hardly noticeable amongst the immeasurable riches of the planet. So we gave little thought to the destruction, only looking ahead to a horizon we never seemed to reach. However, some measure of what we were doing tens of thousands of years ago can be understood from an event that left the last and best-preserved record of the effect the axe may have had in Eden. The event took place only a thousand years ago, when the Maoris arrived in New Zealand.

At the time, the dominant animal there was the moa, a large, flightless bird. It weighed anywhere between 10 and 200 kilos, and as there were no threatening mammals it had taken over the role normal to browsers and fruit eaters. Moas were so numerous that later European settlers often found ploughing difficult because of the sheer number of their bones. But within five hundred years of the Maori arrival, every moa in New Zealand had vanished. Archeological evidence shows that moa meat had been so plentiful it supported the first Maoris in groups as large as fifty without the need for agriculture. The Maoris took the moa to be a free lunch and only later learned there was no such thing.

The Maoris also burned large tracts of vegetation that had very few fire-resistant plants, so within a few hundred years there were parts of the country that the settlers had transformed from a rich, diverse ecology into a virtual desert. Only bracken flourished, because it was resistant to fire and did not need the moa to spread its seeds. Having originally enjoyed a rich and varied diet, the Maori settlers were eventually reduced to surviving off bracken roots.

The Maori destruction of their own environment in New Zealand was only the last in a lengthy catalogue of destruction by Stone Age cultures. Easter Island, now bleak and windswept, was once a forested, subtropical island. In Australia, the devastation was even more dramatic after the continent was first settled around 60,000 years ago. Since then, an estimated 86 percent of the mega-fauna (animals weighing over 44 kilograms) have become extinct. Not all archaeologists agree, but there is considerable evidence that it was humans who wiped out at least 43 large vertebrate species, including a six-foot-tall, one-ton kangaroo, as well as a rhinoceros-sized wombat and a 100-kilo emu. These and very many other local losses were relatively unimportant to humans in prehistory, because their short-term advantages far outweighed any negative ecological effects which, in any case, would have been immeasurably too long-term to matter to anybody in a small group in a big wilderness in any foreseeable future.

As time passed, the ready acceptance of the gifts became a habit that in turn taught us to hold in special regard those who offered them. Axes also conferred power on those few among us who were able to use them to command the community through myth and magic, or their later equivalent, science and technology. Because of this, through history, axemakers have been encouraged by those in power to innovate for them, unhindered. This act of self-interest on the part of the institutions and authorities has generally been disguised by the careful definition of axemaker activity as "free inquiry."

In particular, the gifts made it easier for leaders to extend their power to shape and direct their communities through the increasing command of information. Those who knew how to use the Montgaudier baton, the commodity tokens of the Iranian mountains, the cuneiform of Mesopotamia, the alphabet of Greece, the moveable type of Gutenberg, or the symbols of math, medicine, and science were generally free to act as if the Earth were still as limitless as it had been for the ancient axemakers of Africa.

In virtually every case, though, their use of axemaker knowledge brought immediate benefit to the community, which in turn abdicated responsibility and power to them so long as immediate survival and a rising standard of living were assured in return. This chapter looks at the way in which those attractive, short-term gifts have generated unattractive, long-term problems because of the way, as innovations proliferate, they interact and cause unexpected effects.

The brief catalogue of calamity that follows in this chapter reveals the extent to which the axemakers' gifts have given us the rope with which to hang ourselves. The gifts themselves did nothing to cause an immediate harm or alarm. But the acceptance of each gift caused a change in the way humans saw their relationship with each other and with nature. Each time, in accepting and using the gifts, society deliberately or heedlessly shucked off old values and adopted new ones. Each time, the extra constraint on our behavior required by the use of the new gift seemed relatively minor and in any case well worth it, for the immediate benefits involved. But cumulatively, the effects would be severe.

Twelve thousand years ago agriculture was a quick-fix answer to a short-term problem. When our ancestors first stopped foraging and settled in the first tiny villages, axemakers gave the new communities the gift of survival through techniques for cultivating food. A small number of local cereal plants became our staple diet, and careful tilling of damp soil produced harvests that ensured the survival of the new communities. In time, harvests became big enough to provide surplus, and as a result population numbers began to climb.

From then on through history, axemakers boosted the carrying capacity of the land with ever-more sophisticated and productive tricks: Egyptian irrigation secured crop water when the rains failed; Roman wheeled ploughs cut the heavy soils in the rich lands of the north and made them arable; new fodder crops in seventeenth-century Holland nitrogenated the soil and tripled yields; a hundred years later chemistry provided fertilizers to help feed the soaring population of the Industrial Revolution.

By the 1950s, the response to demand brought innovations in food-producing techniques that would have unexpected and wide-ranging side-effects. At the time, millions of people in the Third World were saved from death by the latest specialist miracle, the so-called "Green Revolution." This introduced new, high-yield varieties of grain and rice and made possible spectacular increases in food production. The new strains solved a variety of local problems because of their relative insensitivity to differences in soil and climate, both of which are extremely varied in semitropical areas.

In the case of rice, under most conditions the new strain produced higher yields because the breeders had used a dwarfing gene from rice plants in Japan and America to produce a short, sturdy stalk on which standard agronomic practices could grow a much bigger ear. With massive fertilization and plentiful water, this meant a two-thirds increase in total biological gain. It looked as if good times were here at last.

The catch was that the technique was an axemaker's gift, so if it were to be successful, indigenous agricultural practices would have to give way to it. So the alternative of increasing yield by traditional methods was dismissed as being insufficient for the short term. Development of "intermediate technology," one adapted to local circumstances, would not enhance productivity as much, but would maintain labor intensity and keep local farmers independent of imports. But local circumstances were ignored, for similar short-term reasons. No attempt was made to concentrate on keeping and reinforcing the balanced, traditional local diet of grains and legumes. Acceptance of the new techniques was seen as the acceptance of modernity.

So any local socio-political objections that might have inhibited the acceptance of the new techniques were classified as "impediments to progress." In 1966, President Johnson of the United States refused India more than one month's drought-relief funding until that country had agreed to adopt the full Green Revolution package.

Earlier in the decade, after nine years' local work on the development of new rice strains, international pressure forced the dismissal of the Indian research director when he resisted handing over rice germ-plasm to the new Green Revolution research institute in the Philippines and publicly urged restraint in adopting the new varieties. But by and large, there was little if any opposition to the scheme from local growers, whose interests were typically (and not unnaturally) driven by the need to make money and support their families.

In the 1960s, techniques similar to those used on rice produced a superior strain of wheat, which was then planted on most Mexican wheatfields and doubled their yields. By 1967, the new rice strains had been introduced into Indonesia. By 1965, the Mexican wheat was in Pakistan, and by 1970 it accounted for half the country's wheat production. From 1965 to 1970, India used new rice and wheat strains to raise production by more than half, and by 1985 the new supercrops were providing more than half the rice and wheat in the Third World (in particular, over three quarters of Latin American grain and nearly all of China's rice). Between 1965 and 1980, global production of these two crops went up by three quarters and the land under cultivation for them rose by 20 percent.

The difficulty with the Green Revolution strains was that it ignored Third-World growers' experience of their local environment and made them instead rely on high levels of fertilizer input from First-World sources. From 1965 to 1975, fertilizer use accounted for over half the yield increase in the developing countries, bearing out the 1967 claim of Norman Borlaug (who in 1970 was to receive the Nobel Prize for his work on the new strains): "There is no more vital message than this: fertilizers will give [India] more food."

Between 1970 and 1973, the trap was sprung. Thanks to a major increase in the price of oil, fertilizer prices increased fourfold, while in the same period Green Revolution agricultural output only doubled. The same occurred again in 1980. Meanwhile, from 1950 to 1975, the area of cultivated land worldwide only rose by a fifth, while global use of expensive fertilizer increased seven times.

The new plant varieties also needed more care and attention than traditional strains. Growing them required more machinery, more capital investment, and, above all, more foreign exchange to pay for all the hidden extras. The long-term social effects in India meant that areas of the country were selected for the introduction of the new plants because of their plentiful supply of water, already developed large-scale farming infrastructure, and the ready availability of skill and capital, so production concentrated in the hands of the big farmers. Then, when prices of oil-based fertilizer rose, small farmers sold out, so the big farms got bigger. The unemployed laborers released onto the market then moved to the new growing areas and their arrival heightened social tensions.

The Green Revolution was in many cases so spectacularly successful that it radically reduced the number of local alternative crop varieties and left the Third World dependent on a few monocultures vulnerable to pest and disease. In many rural areas, those who had not benefited from the Green Revolution, because water supplies were scarce or irregular, left the low-yield land to find only unemployment in the cities. Above all, the Green Revolution put the Third World in debt to the West to pay for huge external sources of energy, research, materials, shipping, processing, marketing, and machinery.

Today global agriculture may be in a precarious state because, thanks to spectacular successes in scientific agriculture, 90 percent of the world's food is now produced from only eight species of livestock and fifteen species of plant. Consequently, the genetic base from which future alternate food sources might come has been drastically reduced, as has the traditional knowledge-base that might have provided alternative techniques one rainy day. Once again short-term gain is paid for in long-term consequences.

Five thousand years ago another life-saving gift that changed our attitudes was the technique of irrigation. When the first irrigation systems were developed in China, Egypt, Mesopotamia, and the Indus Valley, they and their management techniques fed and organized rising population concentrations in large villages and triggered the beginning of civilization.

Since then, engineers have ensured ever-increasing water supplies with water-lifting "shadufs" (ancient Egyptian scoops that used counterweights); dams and channeling supplied a full range of medieval water-powered machinery; suction pumps brought up water in the Renaissance; today's wells go deeper than ever before thanks to carborundum drill bits; enormous volumes of reservoir water are available for hydro-electricity production or irrigation; and rivers have even been diverted to supply entire countries. And every time, the techniques have materially improved life for a growing number of people.

But in the twentieth century, demand for water has increased to such an extent that the perennial sources are now being used up faster than nature can replenish. As a result, water is perhaps the most deficient resource on Earth. One-third of the entire global food crop comes from the 17 percent of the world's irrigated cropland. In China, Japan, India, Indonesia, Israel, Korea, Pakistan, and Peru, irrigated land provides half the national domestic product. Seventy percent of the fresh water taken from the global hydrological cycle is used by agriculture and is essential to the 35 percent of the global land area that would be desert without it. Total irrigated land area has increased by a third in the last twenty years.

The forecast is not good. Global domestic and industrial water use has quadrupled since 1950, and by 2000 demand is expected to double, most of the extra being taken up by the developing countries. By 2000, water shortages are expected to affect 450 of the 644 cities in China where, in the North China plain, demand already outstrips supply. Scarcity is now common in 26 countries, including Russia, the Middle East, parts of India, Africa, and the Southwestern U.S. In the American Great Plains, the giant Ogallala underground acquifer, filled during the last Ice Age, is being drawn down at the rate of six feet a year. That is twelve times faster than replenishment rates. Meanwhile, pricing policies that do not include social or long-term replacement costs invite the average American to use four times the amount of water required by the average Swiss and seventy times that by a Ghanaian.

More than 70,000 years ago specialist axemaker gifts ensured the survival of those who lived on coastlines with harpoons, hooks, and nets for fishing. Over the centuries since then we have taken food from the sea in ever-greater amounts, thanks to better ships, safer navigation, weather forecasting, radar, and the many other industrial innovations, which have made it easier to plough the oceans for their rich harvests. We have also used the oceans as a dumping ground for waste generated by other industrial techniques.

Today, in consequence, fish harvests are failing and the oceans are dying. According to the U.N. Food and Agricultural Organization, four out of the world's seventeen fishing zones are already overexploited. Between 1950 and 1990, the catch increased fourfold, and many believe the current world catch has reached the limits of sustainability.

Contamination is also causing major salt- and fresh-water problems. After the construction of the Aswan dam in Egypt in 1965, natural river discharge from the Nile virtually ceased, with catastrophic effect on the southeastern Mediterranean fisheries. In the Chesapeake Bay, one of the world's richest estuaries, pollution cut the oyster-bed output from eight million bushels per year a century ago to one million bushels in the 1980s.

Human activities involving deforestation, mining, dredging, and erosion also cause sedimentation that fills reservoirs, lakes, and rivers. Sediment containing nutrients such as fertilizers cause massive algae bloom and in turn high levels of fish death. Sedimentation also triggers a fatal condition in coral reefs called "bleaching." This was first seen in 1987 in the Caribbean, but now it appears worldwide, most severely in Asia. Although the reefs cover only 0.17 percent of the ocean floor, they support wide biological diversity, and the food they provide may sustain a quarter of all fish types in the developing world. At current rates of destruction, it is estimated that another two-thirds of the world's reefs will be lost in the next forty years.

Six hundred thousand years ago, one of the first great gifts to follow the axe was that of fire. Heat kept winter at bay and introduced the magic of cooked food. So for millennia afterwards, the woods were unconcernedly stripped for fuel, because if the traveling tribes ever happened to return to the same spot, the forest would always have replenished itself.

By the time of the Roman Empire, fuel wood was in scarce supply, and by medieval times in Europe people were already digging open-cast coal mines for an alternative fuel. Then in the nineteenth century physicists and chemists like Nicola Tesla and Benjamin Silliman saved the day with electricity generators and petroleum, both of which offered apparently unlimited supplies of energy. The sky, it appeared once more, was the limit.

Today we know better, but in the meantime the effect of uncontrolled energy use has been globally devastating as consumption has grown with virtual disregard for the consequences. The United States uses a quarter of global energy supplies, although it has only a twentieth of the world's population, while the Third World, with over three-quarters of the world's population, has quadrupled its energy use since 1960. By 1991, it was using a quarter of the world's coal, oil, natural gas, and electricity output. If present rates of global consumption continue, oil reserves are expected to last only fifty years and gas will last only two centuries. Coal will last about 3,000 years, but very few Third World countries have any of it.

Back in the eighteenth century, the gift of mass-production systems from Watt, Papin, Priestley, Lavoisier, and others caused the surge of industrial innovation that would trigger today's massive energy use. The Industrial Revolution happened in order to satisfy the rapidly increasing demand for goods among a fast-growing population, who were eating well and having more children because of earlier improvements in agricultural output.

All over Europe, water power gave way to steam, and as a result manufacturers were able to locate their production lines on the coal fields and then to move their workers to the factories. Thousands of farm laborers flocked from the countryside to the new cities for all the benefits that a regular cash wage would offer, as ownership of farms became concentrated among fewer people and rural unemployment rose. In any case, life on the farm was no picnic, and industrial cities were magic places where people could get rich quick. Alas the streets were not paved with gold, in spite of Irish ballads to the contrary. In the short term, living standards rose dramatically, but soon the overcrowded and unsanitary conditions prevailing in the industrial towns proved fatally attractive to the cholera bacillus. Hundreds of thousands died from what their new industrial, wage-earning lifestyle had brought them.

Today the axemakers have solved many of those health problems they created with industrialization, but the effects of mass production can be seen in the levels of pollution threatening every part of the planet. Globally, 1990 estimates indicated the annual release into the atmosphere of nine million tons of sulfur dioxide, 68 million tons of nitrogen dioxide, 57 million tons of suspended particulates, and 177 million tons of carbon dioxide. Airborne pollution is now definitely linked to increases in the incidence of heart and lung diseases and, near heavily industrialized areas, pollution also causes large-scale damage to vegetation and wild life. Pollutants generated by traffic are currently concentrated in the industrialized countries, where, since 1950, the vehicle fleet has increased tenfold and is predicted to double in the next twenty years.

National air-quality danger thresholds are frequently exceeded in many major cities. In the case of Los Angeles, Tokyo, and Mexico City, concentrations regularly exceed four times World Health Organization guidelines. Mexico City is a particularly extreme case, where five and a half million tons of contaminants are released to the atmosphere from 36,000 factories and more than three million cars. In 1988, over half of all newborn babies in the city had enough lead in their bloodstream to cause neurological and motor impairment. A 1992 study revealed that globally the average number of nanograms of lead absorbed into the adult bloodstream daily between Paleolithic times and today has increased in air from 0.3 to 6,400, in water from 2 to 1,500 and in food from 210 to 21,000. All in the name of progress.

WHO reports that over six hundred million people are exposed to dangerous levels of sulfur dioxide from fossil fuel burning. One estimate puts U.S. health care costs due to pollution and lost productivity at $40 billion. Russia is a real basketcase. In 1993, studies revealed that one in ten infants suffered birth defects, and more than half of school-age children had pollution-related health problems. Illness and premature death rates were also rising in the 25–40-year age group, and life expectancy was declining.

Industrial waste dumping since 1993 has also contributed to the general pollution. A hundred tons of toxic heavy metals and thirty tons of toxic chemicals from the Rhine pour into the North Sea each day, while since 1985 levels of tritium entering the same sea from the River Meuse have risen by three-quarters. An estimated 100 square kilometers of the Baltic Sea is an ecological wasteland with no life below 80 meters' depth, making it one of the most polluted seas in the world, thanks primarily to effluent from factories, farms, and households. The Mediterranean takes an annual total of twelve million tons of organic waste, 320,000 tons of phosphorous, 800,000 tons of nitrogen, 100 tons of mercury, 38,000 tons of lead, 2,400 tons of chromium, 21,000 tons of zinc, and 60,000 tons of detergents. Thanks to the original gift that made possible the mass production of mass possessions, we now live in a dangerously degraded environment.

In terms of what we might call the "axe-mark," from prehistoric times each individual new tool left its individual, short-term legacy in the form of local areas of forest laid waste by slash-and-burn or stripped for firewood, or else a few dozen square miles of land spoiled by coal-mine slag heaps, a river darkened by factory effluents, or a small town blackened by smoke from factory chimneys. But the modern world is now so interconnected that these local effects have begun to form global patterns of destruction that cut into the most fundamental sustaining resource on Earth now being rapidly degraded by the effects of centuries of ingenuity: the land.

It takes a period of 3,000 years for natural processes to replace the six inches of topsoil necessary for crop cultivation. One inch of soil loss can reduce corn and wheat yield by 6 percent. Every year six million hectares of arable land are so degraded by soil loss they have to be abandoned. Erosion and soil degradation occur mainly because of the effect of wind on land after ploughing, mining, or the flooding that follows removal of forest cover. This is now affecting up to half of all land on Earth, reducing the productive capacity of arable areas by between 25 percent and 100 percent. One-fifth of all cropland has been irretrievably lost through degradation, as well as a quarter of once biologically productive but as yet uncultivated acreage. Since 1984, the total area going over to crops has actually stopped growing. But the rise in the number of mouths to feed has not.

Removal of topsoil since 1972 is reckoned to be around five

hundred billion tons, a rate of loss perhaps forty times faster than the natural replacement cycle. In India alone, the amount is annually equivalent to the surface area of the state of Massachusetts. Loss also occurs because of desertification, lack of water, erosion, and forest clearing, all of which annually keep twenty-one million hectares of land (on which 850 million people live) too dry to use. The thirty-fold increase in the area of irrigated land in the last two centuries has also generated major salinization problems, as has the increasing draw-down of freshwater sources. The land is the ultimate resource and we are fast losing it.

Two million years ago, the earliest of all gifts to improve the fortunes of humankind was, of course, the axe to which this book is dedicated. With the axe we made the wooden artifacts that would help us survive, including carpentered shelters, weapons and tool hafts, and handles of all kinds. As the population grew (and with it the need for more of everything), wood became the prime material for construction. And once the axemakers had provided means to build and navigate them, the second major use of wood was for boats and rafts.

For most of history, overland travel was impossible in bad weather for months at a time, so most commercial activity up to modern times took the maritime route. So it is hardly surprising that centers of population developed along coastlines and that soon these began to suffer from wood famine. Governmental edicts from the time of ancient Greece already point to a growing concern with forest loss. By the seventeenth century, when the construction of a navy man-of-war needed a thousand oak trees, there were official bans throughout Europe on the use of forests for any other purpose.

But these minor local historical difficulties pale into insignificance beside today's problems, as Earth's forests disappear before the relentless chainsaw. Since 1700, the average area lost each year has been equal to the size of Switzerland, but more than half the total loss has occurred since 1950. In the past decade alone, the forest loss has equaled the size of Malaysia, the Philippines, Ghana, the Congo, and Ecuador combined. Best estimates are that Africa has already lost nearly three-quarters of its original climax forests and Asia two-thirds.

In heavily forested countries like Brazil, unregulated slash-and-burn agricultural expansion causes major clearance and so does the opening-up of forest land for resettlement. This is happening on a major scale in India, Brazil, and Java. Many developing countries with major debt repayment problems have also resorted to selling their forests.

With rising populations and limited indigenous energy resources, Third World countries are also stripping the forests for fuelwood because they cannot afford to import coal or oil or gas. A vicious circle develops: with insufficient gains in agricultural productivity, due either to inequitable land distribution or to the introduction of high-cost Western farming methods, increased land clearance is necessary in order to feed the population. This reduces agricultural and forest productivity, and that in turn triggers the need for more forest clearance, with devastating long-term effects. Because of this kind of deforestation in the Himalayan watershed catchment areas, five hundred million Indian and Bangladeshi small farmers now live in a flood-or-drought regime.

Also, as the forests disappear, whole ecosystems are being cut off into isolated biological "islands," causing floral and faunal collapse as their vital links with other ecosystems are severed. The total number of species in existence is unknown, but it may be as many as thirty million, one-third of which live in the tropical forests. In the Peruvian rainforest, for example, one tree type is home to 43 species of ant. Biological diversity like this is being drastically reduced by mass extinctions triggered everywhere by pollution, acid rain, flooding, land degradation, and deforestation.

And as the human race continues to wipe out life-forms in this way, the interactive nature of the ecosphere makes it difficult to assess the overall effect. Some estimates put the loss rate as high as ten thousand times faster than before the advent of humans. If this is true, then we are losing species at unprecedented speed, with vertebrate and bird species reckoned to be disappearing at between 100 and 1,000 times the natural rate. We go on doing it, even though we don't know whether what is happening is catastrophic or just Armageddon.

The major danger associated with the loss of tropical forests is that they have historically been the source of genetic replenishment after natural extinctions, which tended to happen at an average rate of a few species every million years. At present logging rates, additional tropical forest loss in the next thirty years could be as much as 15 percent. If that happens, then South American forest loss alone could account for the disappearance of 15 percent of all plant species and up to 70 percent of birds. Rachel Carson's *Silent Spring* will become a reality.

As has been said, perhaps the most influential and pervasive of all effects on the planet of the axemakers' gifts was caused by the Industrial Revolution. From the early nineteenth century it brought rapidly rising standards of living and possessions for all. But with the massive surge in fossil-fuel use for fuel and electricity generation that accompanied the Revolution, the Earth's atmosphere also began, imperceptibly, to change, as fossil-fuel burning generated more and more carbon dioxide.

This gas (and other gases such as methane) act to trap solar energy in the atmosphere and prevent it from radiating back into space. When this happens, the atmosphere warms up in a process known as the "Greenhouse Effect." If today we were to stop all emissions of these gases, most experts believe that the amount of increased heat from the greenhouse effect would still cause a mean global temperature increase of between four and seven degrees Fahrenheit.

But the ecosystem is so interactive that feedback mechanisms could powerfully enhance the effect of this apparently innocuous amount of extra warmth. But in any case, it is not so innocuous as it seems. The following events were generated in the past by a mean global temperature rise of only 7°F: the descent of humans from the trees over thirteen million years ago; the last Ice Age 120,000 years ago; the colonizing of America from prehistoric Siberia around 18,000 years ago; and the rise of civilization in the Middle East, India, and China 6,000 years ago.

In the recent past, the contribution of fossil-fuel-generated CO_2 has come mainly from the industrialized countries (over two-thirds of global emissions in 1990), but if Third World populations continue to grow at the present rate, their emissions will more than double by 2025, boosting total global totals by half. Overall, greenhouse gases are forecast to double by 2025.

Greenhouse could have a number of effects that should concern everybody. Melting of the polar icecaps would cause a rise in sea levels and major forests would die due to the inability of plant species to adapt to high rates of temperature increase. The same would be true of crop yields. A small increase in sea temperature would cause more powerful and frequent major storms in Southeast Asia, Australia, the Caribbean, and the East Coast of the U.S. Global warming could also cause dieback of the oceanic phytoplankton that absorbs three billion tons of carbon a year from the atmosphere, leaving the carbon in the air to further enhance the warming effect. And a sea-level rise would cause major population displacement and death among the two-thirds of the world's population that lives on low coastal land. Again, we don't know exactly how bad things will be, but we go on doing it.

All these causes and effects of uncontrolled cut-and-control that we have indulged in throughout history have brought depressing results. But in every case one factor is consistently present, making things even worse: population growth. Ironically, the rise in the number of humans was until recently considered to be the greatest of all successes because it signaled more and more effective ways to combat death by starvation and disease. At any time in the past, larger, healthier, populations that were increasingly well fed, clothed, and housed were the most convincing argument for taking everything the axemakers offered.

But in the case of population growth, thanks to the increasingly interactive nature of the world, the gifts have worked too well. After the early benefits of food, warmth, and shelter, survival rates were improving significantly even in primitive times, thanks to knowledge of the medicinal value of plants. By the time of the Greeks, a pharmacopoeia of various natural remedies was curing ailments and saving lives. At the time of the Renaissance, new surgical techniques were being developed, together with a general advance in the knowledge of anatomy.

The spectacular bacteriological advances of the early twentieth century conquered major respiratory diseases, as well as typhoid, smallpox, diphtheria, malaria, and most parasitic diseases. During World War II, new public health techniques were implemented on a major scale in order to protect the health of Allied troops, especially those serving in the tropics and in the Far East. The problem of malaria and yellow fever was tackled by means of mass immunization and vaccination, as well as the introduction of large-scale spraying of potential malaria sites. Western technology was also introduced to drain tropical swampy parasite breeding grounds, and epidemic control was achieved above all through the mass use of DDT.

After World War II, the spread of these techniques brought changes that have since made life both dramatically better and disastrously worse. The mortality rate in Third World countries dropped like a stone as plagues and pestilence were virtually eliminated within a generation. The world was suddenly much healthier. This piece of good news was followed by a sudden and spectacular increase in population growth.

Inevitably, as a result, the human race today crowds in ever larger groups. In 1950, Mexico City contained 3.1 million inhabitants, but by 1989 the number had increased to nineteen million and a population density four times that of London. Calcutta grew from 4.5 million in 1950 to over 10 million in 1994. Most Asian countries expect a doubling of urban populations by 2010, and by 2025 there will be 639 cities of over one million people, 486 of them in the Third World. None of these cities will be capable of supporting their populations or of providing water, sewage, and garbage disposal for their populations, who will live in abject poverty and disease. So much for good news.

Today one hundred million people on the planet are homeless, four hundred and ninety million people suffer from severe malnutrition, eight hundred million are illiterate, and three hundred and sixty million have no jobs. The 1992 standard of absolute poverty as an annual income of under $500 a year fits seven hundred and fifty million people, most of them living in Africa. Innovation with little or no regard for long-term social effect has brought human numbers close to overwhelming the planet. By 2000, 40 percent of the population will face severe land shortage and 1.2 billion will risk starvation. Meanwhile the U.S. spends $5 billion a year on slimming diets, and the average citizen of an industrialized country consumes twenty times the resources used by a member of the Third World.

Population forecasts are catastrophic because even if the world's birth rates have indeed slowed in recent years, the one billion humans still under fifteen years old who will grow up and have children means that the population will likely more than double to 11 billion by the year 2060. If it takes until 2040 to reduce the average Third World family to two children, the global population in 2100 could be as high as 15.3 billion.

That population will be very different in composition from what it is today. According to the World Resources Institute, by 2050, 84 percent of humanity will live in what is now the Third World and half of them in only five countries. These shifts in population density and distribution will have serious implications for food, resource distribution, and the political makeup of the planet. In 1994, the United States was the only industrialized democracy in the five most populous countries, but by 2025 all the present industrialized democracies will be "small-number" nations. The U.S. population will be less than that of Nigeria, Iran's will be twice the size of Japan, and the population in Canada will be smaller than Madagascar or Syria.

All in all, the fate of the axed Earth seems to be dictated by Stanford population biologist Paul Ehrlich's equation: "Impact on the biosphere = size of population × resource use × technological effects on the environment." At the moment, a million new humans will now be born every five days. So what consequences will Ehrlich's equation bring and how can they be dealt with?

In 1798, an obscure English cleric and amateur axemaker called Malthus wrote several pessimistic pamphlets on population. For Malthus, the fundamental problem was that while populations increased geometrically (1, 2, 4, 8, 16, 32), food production only increased arithmetically (1, 2, 3, 4, 5, 6, 7). So the human race would always inevitably outstrip its food supplies. However since this rarely seemed to happen, Malthus reckoned there had to be a limiting mechanism at work.

When, as Malthus believed, the population increased above what could be fed by the land, the most effective check on further growth was usually death. But this kind of geometric population growth occurred so rarely that Malthus came to the conclusion there were other factors at work. One he called "preventive," acting in time of shortage, when people deferred marriage and child-bearing. The other he named "positive," when the checks were the result of war or disease or famine, when rising mortality rates would "positively" reduce population levels.

But Malthus lived before the Industrial Revolution, when chemical fertilizers and farming technology artificially enhanced the productive capacity of land and provided adequate food supplies for rising populations. The result was dramatically different from anything Malthus might have forecast, thanks to a phenomenon we went through, called "demographic transition."

For most of history, high birth rates had been maintained through religious practices, moral codes, laws, educational practices, marriage habits, or family structures, and these were traditionally countered by equally high mortality rates due to disease, famine, war, and epidemics. Then industrialization tipped the balance in favor of population growth. Improved public-health facilities and increasingly reliable food and water supplies caused the mortality rate to fall and life expectancy to increase. Under these conditions, population growth accelerated.

At this point, around the mid-eighteenth century, urbanization and industrialization began to affect the fertility rate. A rising standard of living, better employment opportunities, higher wages, and increased savings made the labor value of children less important. Then the provision of schooling removed children from the job market. Medicine reduced the infant mortality rate, so more children survived to adulthood and families could afford to be smaller. Overall, through the early twentieth century as the economy expanded and more employment made more people better off, rising income tended to lead to falling populations. At the same time, labor-saving machinery generated surplus labor, so by 1975 the population growth rate of the industrialized countries had dropped close to zero.

What can be done to help the twenty-first-century Third World achieve the same demographic transition before it succumbs to mass starvation? Given that the present global crisis described in this chapter has been caused by the long-term effects of short-term quick fixes, the solution to the problem is unlikely to be found through reliance on tried-and-true strategies from the past. The old answer, a massive push to industrialize the entire world, is manifestly out of the question. But in this and in all the other difficulties the axemaker gifts have created, it is not going to be easy for voters in any country to decide how to escape from the situation, because they know too little. Over the centuries, the majority of the community have been excluded from the increasingly specialist information their leaders and institutions have drawn on in making policy decisions.

In these modern times, when it is fashionable to call information the "commodity of the future," it is perhaps worth recalling the way in which information has always been used as an artificial "possession" by those who had it. At the time *Australopithecus* was fashioning the first stone implement (and with it the means to make things that had not existed before), the hominids' relationship with the environment was so interdependent that they might be said to have been an integral part of it. Then the axe brought into existence something that was separate from this shared reality. Thanks to the axe and increasingly because of the gifts that followed it, there would now be a new, unnatural element in life. This entity would become known as "knowledge," and it would serve to change the world and to control people.

Over the centuries, knowledge proliferated and took such widely differing forms as bone batons, irrigation systems, alphabets, logic, numbers, theology, experimentation, steam engines, post-graduate degrees, or bureaucracies. At any time only a small fraction of any particular community had access to these instruments of change, and the most effective advances in knowledge were those which made possible the manufacture of more like them. So external memory storage and communication devices like tokens, letters and numbers, papyrus, print, telegraph and radio all triggered surges of innovation that strengthened the position of those in power.

The trick of the matter each time seemed to be that those with knowledge could create magic things that only the possessors of knowledge could either see or use. These were intellectual worlds accessible only to people with the ability to read and write, use syllogisms, look through telescopes or down microscopes, make radio signals, or smash atoms. It was not until the twentieth century that many members of the global community even knew these worlds existed, let alone came to grips with understanding them.

The social effects of the axemaker's gifts have all too often wrong-footed even society's institutions because the way innovation triggers more innovation has made life much more complex. Straightforwardly simple artifacts seem to cause the most unexpected multiple outcomes. For example, printing made possible the production of maps that could be regularly corrected and reissued. In order to do this in seventeenth-century Holland, ships' captains returning from voyages were legally obliged to hand over their log books and navigation papers and were also subjected to detailed questioning at the quayside.

This, in turn, made possible up-to-date, standardized cartographic data that were then made available to other merchant venturers so as to make their voyage that bit less dangerous. This made the business of exploration easier and so attracted investors not unnaturally interested in the extremely high levels of potential profit. So investment in voyages increased and, in an effort to generate more cash for potential backers, the English invented a Land Register, which made it easier to use land as collateral to borrow money from the new mortgage companies set up for that purpose.

While some of the borrowed money went to pay for ships or to underwrite voyages, more went into the new insurance business, recently set up to take the risk out of the voyages. Risk was reduced even further by the invention of the limited company, trading within another new entity, the Stock Market, that in turn needed the kind of stable financial management only a new, national bank could provide. The first bank was, not surprisingly, established in Holland. Data from the maps and Land Registers and insurance contracts and bank accounts eventually gave governments control over the financial affairs of their entire countries because they made possible new mercantilist policies based on these means of monitoring commercial activity. And all of this came from the printing of maps.

Even if extraordinary forethought might have made it possible to second-guess the link between maps, exploration, insurance, stocks, and finance in this particular case, in general the complexifying effect of innovation is too accidental to do so. For instance, Venturi's eighteenth-century theory about water pressure led, through the application of the theory in the nineteenth century antiseptic spray, to the modern carburetor that would help to generate the American way of life made possible by the automobile. In another example, in 1912 the English meteorologist C.T.R. Wilson's interest in rainbows led him to study them in the laboratory with an artificial cloud-making machine, which in time revealed the existence of sub-atomic particles that made possible the explosion at Hiroshima.

As we've seen in this chapter, the problems the world currently faces are the result of many such serendipitous events, each of which brought immediate benefits but all of which also interacted over time to generate complex effects that could never have been expected. If society is to find long-term answers to the critical problems that as a result it faces today, we need to find a way to escape from the short-term view natural to our most primitive instincts (and reinforced by millennia of quick-fix gifts we could never refuse). We also urgently need to find a way to control the process of change itself.

Is there any simple resolution? Probably not. The world is too interdependent for quick and easy fixes any more. The best that can be hoped for is to find some way of assessing what limited options are left to us, as well as the potential long-term effects of whatever decisions we take. There is no single solution to the multiple problems we face. The axemakers' gifts have made life too complicated for that.

Ironically, given the overwhelming importance of population growth in all this, the first step toward an answer may come from a tool invented to manage the massive population rise in America at the end of the last century. The new tool was a quintessential axemaker gift because it made easier and more effective than ever before the manufacture of knowledge and the cut-and-control of society. It is also ironic that the same system might ultimately provide a solution to the myriad problems facing us today precisely because it may make possible a radical change in our relationship with those same axemakers.

Chapter 11

FORWARD TO THE PAST

Human beings are the only species with a history. Whether they also have a future is not so obvious. The answer will lie in the prospects for popular movements, with firm roots among all sectors of the population, dedicated to values that are suppressed or driven to the margins within the existing social and political order: community, solidarity, concern for a fragile environment that will have to sustain future generations, creative work under voluntary control, independent thought, and true democratic participation in varied aspects of life.

NOAM CHOMSKY

As this book has tried to document, the future faces us with serious difficulties we have inherited from the distant past.

Millennia ago, before the appearance of the axe, the day-to-day existence of the primeval individual extended for perhaps dozens of miles, rather than continental distances. This is still the case for most of us. Thirty thousand years ago, upper Paleolithic people knew at most a group of between fifty and two hundred individuals. This is also still the case.

Up to 30,000 years ago, human mental processes had evolved in large part to deal with immediate problems: deciding which berries to eat, how to survive the winter, how to avoid dangerous animals, and when to find shelter. These were the mechanisms with which blind evolution had prepared us to handle the world. Our mental predispositions, like that of every other animal, were circumscribed by the immediate horizon and by short-term problems. This was in every sense "natural," because there would have been little point in worrying about the long term if immediate threats such as tigers and winter were not dealt with.

Our ancestors also never had to deal with all of humanity as a factor in their daily lives, because for most of history they only knew a small number of individuals going about their particular activities in a very small world. They could slash-and-burn a forest or wipe out many species and then move on, because there was plenty of Earth and few of us. There was never the need to consider the entire planet because it was too big for us to have any meaningful impact.

Today, however, humanity is no longer made up of a few scattered bands of people spread out thinly on the planet. Humanity is now a monster, producing many more people in a month than were alive just before the first agricultural revolution and in aggregate weighing more than any other land species.

Because our lives have changed since those primeval times, it is imperative above all that we revise our out-of-date perception of the world, so that our ancient, small-scale, small-time mind can be expanded to consider more distant horizons and more frequent changes. Many commentators on current problems seem to suggest that a logical or psychological alteration in the way our minds work will, alone, do the trick. But we are mentally so separated from the natural world around us by the axemaker gifts which have, over millennia, shaped every aspect of our lives that both the gifts themselves as well as a change of consciousness need to be parts of the resolution.

The difficulty is that modern human beings no longer directly perceive the world they live in and whose condition affects them. As Chernobyl showed, the world is now too big and too complex for that. And it is not the case that we can somehow throw away all modern technology and return to a simpler, Arcadian life. Even if such a world ever existed and even if some of us could survive in that kind of environment, the vast majority of the population (who cannot farm) would not want to. So we must find a realistic way to handle the problem by tackling it with what tools we already have at hand: available axemaker gifts that might serve for this new task. And since time presses, as we use them we must act short but think long.

The gift with which we might begin to make the necessary change is already in our possession. But we need to be ready for the fact that the effect of its use could bring changes more radical than any so far in our history, because they might take us back to what we were, mentally, before the axemaker's first gift changed the way our minds got developed and selected.

The tool in question first saw the light of day in 1888, in Baltimore, Maryland, when Herman Hollerith, a young engineer, assistant to the Director of Medical Statistics for the 1890 U.S. census, talked to his brother-in-law about the problems that lay ahead for the census takers. The flood of immigrants from Europe had reached almost unmanageable proportions. Ten years earlier, the previous census (of a very much smaller population) had taken over five years to complete. At current growth rates, it might take longer than ten years to count the population every ten. You can see the problem.

Hollerith's brother-in-law worked in the textile industry and he told Herman about the Jacquard silk loom that used a system of perforated paper sheets to control how a pattern was woven. The presence or absence of a hole in the paper permitted or blocked the entry of one or more of a set of spring-backed hooks, each of which could pass through a hole and lift the necessary thread for that stage of the pattern.

Hollerith's adaptation of this idea took the form of cards on which each individual statistic was expressed as a hole. Counting and analysis were done by pressing the card against a set of sprung needles. Where a hole permitted passage, the needle tip would go through to make contact with a live wire on the other side and send a signal to an electric counting machine. Collating the data involved patterned templates that limited which holes were uncovered. In this way, a chosen category, such as first-generation Greek-American grandmothers or Japanese-born carpenters, could be easily and quickly totaled. After the census was over, Hollerith took his counting system to a group of investors who set up a company that eventually became known as "International Business Machines," (abbreviated as IBM). Within fifty years, Hollerith's punched cards had been replaced by keyboards and electronic chips in a machine known as a computer, the greatest of all axemaker tools for cut and control of information.

If the esoteric, unexpected, and interactive nature of the axemaker processes described throughout this book are what have kept institutions and individual leaders in power over a two-million-year period, then the new, knowledge-manufacturing potential of the computer could enhance that power as never before. In the few decades since its invention, the computer has brought change to almost every aspect of modern life and has made society so complex and interdependent as to make the old cut-and-control, reductionist way of thinking too hazardous to operate in its previously isolated and unaccountable way.

The networked computer began, like most technological innovations, as a brilliant, short-term gift designed to solve an immediate problem. The first data terminals linked a chain of North American Arctic radar stations to form the "Dewline" defense system, designed to protect the United States against the U.S.S.R.'s first atomic bombers.

By the 1960s, computers had proved ideal for repetitive, time- and labor-consuming tasks, and they were soon replacing humans in routine clerical work. The machines made information available and processed it on a scale and at a rate never possible before and generated the kind of work that could only be done with the assistance of the computer.

The provision of this new tool for data control changed the way we thought about information and how it could be used, how accessible it might be, how it could aid decision making in every specialist area of activity, and how quickly it would bring change. The effect of the introduction of the computer was to create new kinds of data, bring into existence new kinds of activity, and encourage organizations of all kinds to expand and, above all, to diversify. In other words, computers enhanced the concept that established entities and systems might change more readily and with less risk than before.

The best example of this effect is the first one, when computers left the military arena and entered the civilian world. "Dewline" technology was adapted to become the basis for an airline reservation system known as SABER. At a stroke, this new tool made it possible to run much more complex and extensive airline schedules, and this in turn dramatically changed the world of business.

The computer has compounded its influence by developing faster than any other innovation in history, so it has also altered our time perception regarding how fast rates of change can themselves change. Over the last four decades, data-processing techniques have changed so often and so quickly that they sometimes become obsolete even before they reach the market.

The advances and the numbers involved in the development of computing are so extraordinary they are hard to grasp. Since 1950, the number of circuits per cubic foot has risen from 1,000 to many billions. The time it takes to carry out an operation has been reduced from 300 microseconds to less than five nanoseconds. The cost of processing one million basic instructions has fallen from $280 to less than a tenth of a cent. The number of characters that can be stored has increased from 20,000 to hundreds of thousands of millions. Machines that used to function trouble-free for hours now do so for years. Since 1965, the average number of chip components has doubled every year, and recent advances in molecular-level nanotechnology promise even more dramatic increases. Individual home computers today have more processing power than that available to the Allies in World War II, and their power is increasing yearly.

Early in the computer's development, innovation in communications technology also produced high-speed data networks that would make the computer's work more generally accessible and change our space perception of information, where it could be stored and the extent to which the physical location of data was any longer relevant.

In December 1958, triggered by the 1957 success of the U.S.S.R.'s Sputnik, the U.S. "Score" satellite demonstrated that voice signals could be broadcast from an orbiting transmitter. By 1965, the Comsat organization had set up the first commercial Atlantic satellite, Intelsat 1, carrying either 240 two-way phone circuits for voice and data, or one TV channel. Transnational corporations could now centralize worldwide operations more easily than ever before.

In the 1970s, the converging systems began to cause widespread social change as electronic fund transfer radically altered the nature, range, and function of national and international financial institutions. Large-scale, high-speed transfer of funds between financial institutions and economic organizations at home and abroad undermined the ability of national governments to impose effective exchange control regulations.

As use of credit grew easier in electronic form, personal disposable income became a kind of futures market, in which consumers' potential for earnings represented loan collateral. This ability to exploit the individual's financial future, multiplied a million-fold, began to impact on commercial arrangements and altered Western export-import relations, in turn affecting rates of exchange, the structure of industries worldwide, and international trading and political relationships.

The supercomputer emerged in the early 1970s. Within twenty years, it was able to perform more than a billion calculations a second. Most supercomputers achieve high speeds through parallel processing, which involves networking a number of processors, each one with control over its memory and programmed to do a specific type of task. In this way, parallel processing splits up a job into many parts, all processed simultaneously. The first practical parallel-processing system, The Connection Machine, had 65,000 networked processors.

By the 1980s, governments and private institutions were routinely maintaining millions of files containing personal details of individuals' lives, which in many cases could be sold to private corporations as mailing lists. It is now practically impossible for an individual to leave the system, and with this in mind the potential for data networking raises other questions of privacy. For instance, a minor infraction like a parking ticket might trigger the request for a profile that would put medical, education, and genetic information about the individual into the hands of the police.

On a larger social scale, the effects of the convergence of telecommunications and data-processing since World War II have provided the clearest possible example of the way innovation has tended to make the world more interdependent. As each technological development put more information into the social system, new networks came into existence, linking previously isolated sectors of daily life. In the next century, such networks could make possible highly centralized social management because, as long as the diffusion of information was strictly controlled, they could radically increase the manufacture of specialist knowledge to enable cut and control on an unprecedented scale.

Throughout history, new communications systems have triggered change in the way society could be run (and in the way we thought) because when bits of information come together in new ways, they tend to generate innovation that is more than the sum of the parts. In ancient Mesopotamia, the gift of writing linked astronomy with hydraulic engineering and triggered the beginning of civilization and a new sense of "place." Fifteenth-century printed biblical appendices brought together experts whose shared knowledge set off the birth of "new worlds" of the scientific instrument. The nineteenth-century telegraph united the railroads with the American System of Manufacture, and made the U.S. an industrial superpower and the individual a cog in the machine. The networks of the next century will make it possible to link trillions of bits of information in similarly unforeseeable ways, with effects that can only be guessed at. Many people are concerned that this will further divide society. On balance, we are more sanguine.

As with every major previous advance in knowledge-manufacturing techniques, the computer could also introduce a new factor into the equation. The effect would echo the way early Paleolithic society was directed by bone-carving shamans, Egyptian irrigation systems were organized by bureaucrats, medieval European battle technology could only be used by aristocrats, the sixteenth-century printing press was under the control of civil and religious censors, and nineteenth-century scientific advances, such as the telegraph and the telephone, were controlled by state governments.

In the twenty-first century, a new kind of axemaker will emerge in response to an old problem: innovation has usually been limited by a lack of specialists qualified to manage it, and this has been particularly true in the case of computer technology. But in this particular case, the extra specialists may not be needed, thanks to the recent emergence from the same technology of two new systems.

The first is the electronic "agent," a software program that can act on behalf of the system-user to search for required data, process them in the required manner, and present the results in the required form. The extraordinary speed at which this can happen will make possible the management of radically higher rates of change. And the manufacture of large amounts of knowledge will happen without the need for the community to generate thousands more human specialists (whose qualifications in any case will, thanks to the same accelerating rate of change, be of only temporary value).

The second innovation, the "knowledge-based system," adds to the abilities of, and in some cases replaces, the human specialist. There are several thousand such systems already in use either because there are not enough human specialists available, or because the circumstances are too dangerous for humans to operate. The systems work by simulating some aspects of human reasoning and deduction. Their combination with the "agent" could generate the first electronic axemakers, whose use of supercomputers would be certain quickly to accelerate knowledge manufacture.

There is nothing new about machines that do things that humans cannot. Throughout history, mechanical replacements for humans in processing and communicating information, such as the alphabet, writing, print, and the telegraph, have driven society to evolve, to differentiate, and fragment. This process generated new disciplines, new ways of thinking about the world, and new cut-and-control social hierarchies.

Those structures and communities that were incapable of managing change by adapting to the new processes were on many, if not most, occasions, overtaken by events. In the latest example of this, a kind of electronic colonialism is already reshaping relations between those countries with a computer skill-base and those without. The English nineteenth-century poet Hillaire Belloc described a similar high-tech effect on the "primitives" encountered during nineteenth-century European colonial expansion:

The plain fact is that we have got
The Gatling gun and they have not.

What is different this time, however, is the speed with which computerized knowledge manufacture and its effects are growing and accelerating. In the ancient world, there was only one science—cosmology—from which all other knowledge flowed, but since the time of the Greeks knowledge had developed into scores of esoteric disciplines, increasingly isolated from each other. After the emergence of seventeenth-century reductionism, each major new discipline sooner or later fragmented into dozens, sometimes hundreds, of specialist subdisciplines. Botany, for instance, subdivided and linked with other disciplines to become biology, organic chemistry, histology, embryology, evolutionary biology, physiology, cytology, pathology, bacteriology, urology, ecology, population genetics, and zoology. This process has repeated in many other fields, and the latest count suggests that there are now more than 20,000 separate scientific and technological subjects. Specialists know more and more about less and less, and non-specialists know less and less about more and more.

Specialist knowledge becomes continually more difficult to keep up with because it steadily proliferates and becomes increasingly inaccessible as each new group of specialists develops its own arcane vocabulary in the interest of greater precision. Just glance at any of the more than 195,000 different periodicals now published every year.

The problem has been that because of the esoteric nature of this kind of knowledge, when it has been loosed on the community it has always been presented to people as a *fait accompli* to which they were obliged to react as best they could. Naturally, the community's ability to deal with the situation was in general limited by the systems available at the time, so in many cases the outcome of the surprise was another surprise. The "knock-on" effect has generally been most significant.

For example, the textile machinery that helped factory owners to take England to industrial pre-eminence in the eighteenth century also led, thanks to the speed at which the population of the new industrial towns grew, to living conditions that would bring the country close to prerevolutionary conditions no factory owner would have ever wanted, because the political, educational, and public health systems were outdated by the speed with which circumstances had changed.

The fifteenth-century discovery of America by people who were looking for a quick way to the Spice Islands rendered most of Western knowledge uncertain and totally unbalanced the entire social order because, until that time, authority relied on unquestioned reiteration of ancient classical knowledge and possessed no experimental techniques with which to evaluate new data.

The Copernican proposal of a new, heliocentric solar system, designed to make easier the calculation of Easter, undermined the whole theocratic and political structure of sixteenth-century Europe because, up to then, all social and intellectual systems had been based on the unchanging nature of Aristotelian cosmology.

This problem of adaptation to change has also been made worse by the institutional obsession with keeping knowledge secret. Knowledge gives power to control those people who do not possess it, which is why access to information has so far been limited to those who could pass ritual initiation tests. Monarchs and governments throughout history have never revealed "matters of state," and modern-day corporations and countries operating in the free market are unlikely to tell each other about in-house design or research projects.

Now, computer-generated knowledge has begun to change the world so fast and so surprisingly that the process comes close to outrunning even our basic evolutionary, adaptive abilities. For instance, if two main frame computers were "talking" to one another and somebody used the keyboard at the rate of a fast typist to say to them, "Listen, IBM-8X and Cray-3YY. Can you tell me what you guys are talking about?," in the time it took to type those words two current computers could have exchanged the entire contents of the

Encyclopaedia Britannica more than a hundred times. And the computers are going to get faster.

But if twenty-first-century information technology generates a knowledge base expanding at astronomical speed, will things simply get astronomically worse? If esoteric specialization proliferates, will it fragment society into isolated groups of people unable to communicate with each other because of the increasingly isolated nature of their knowledge? And when the products of machine interaction go public in the form of society-changing innovations at a higher rate and in quantities greater than ever before, how will the general public be informed of what's happening if the media have already forsaken a common forum for hundreds of special-interest cable channels dedicated to mud-wrestling or growing begonias? This is not a distant Buck Rogers possibility. This is today for some and tomorrow for the rest.

A key factor influencing how we approach this problem is our archaic, short-term view of the world, embedded still more deeply in us by centuries of short-term quick-fix innovation. Today, for instance, partly as a result of indoctrination and partly due to our innate tendencies, we find it easy to ignore that the annual cost to the world in emphysema and work lost due to smog is many hundreds of times what it would cost to clean up the smog over a ten-year period. In Mexico City, at present 72 percent of children have brain-damaging lead quantities in their cortex and in their blood, which means that 72 percent of people in Mexico City (and similar numbers in many Third World countries) are going to suffer from limited brain development. This will perpetuate the cycle of undereducated, unproductive Third World people. Does anybody care enough to change? It's not easy to do so, since even in Northern Europe in the last few decades it took fifteen years of campaigning to get only a fraction of the lead out of only some of the fuel.

The same dilatoriness also affects long-term issues, such as excess agricultural production in the first world, the long-term effects of

Third-World borrowing, inner city deprivation, the rebuilding of Eastern bloc economies, psychiatric care in the community, global population growth, use of resources, and many others.

So how do we change the way we think in time to stop short of catastrophe? We have two tools that may help. One is a new tool: the information technology whose potential for harm has just been described. The other is an old one: the brain.

Human beings live all over the planet in the widest possible range of climates, from rain forest to Arctic to desert, eating an enormously varied diet from red meat to brown rice to orange insects, working at astonishingly differing jobs from hair dressing to hunting tigers, or manipulating finances, baking cakes, engineering genes, studying the nature of the universe, and everything in between.

That one species can accomplish so many completely different kinds of self-expression is due to the extraordinary flexibility of the human brain. Although we come into the world with certain abilities, like suckling or looking toward faces, and with talents to be developed, like learning a language and walking on two feet, each of us is also born with many more latent abilities than we can possibly realize in a life span of less than a century. And our brains are, of all mammals, the most unfinished at birth. This feature lets our early environment play a heavy hand in sculpting what we become.

Research on how experience affects the brain is just beginning. In the past, most psychologists and neurophysiologists assumed that our brains were fixed and that we learned to function as well as we could within the limits established by genetics. But many new studies show that the basics of how we operate are changed by experience. This underscores how important it is to evaluate from time to time the ways in which we are changing and reshaping the world, in order better to understand the effect that the changed world might have on us. The classic computer maxim GIGO ("garbage in, garbage out") applies here: you may well be what you eat, but you also are what you see, hear, smell, taste, feel, and do.

What "being flexible" implies is that if we do not like the biases so far put into our brains by the axe-made world, we are not stuck with them. These biases can themselves be changed by the same means with which they were created, because modifications to brains happen frequently and naturally. As we have shown, the acts of writing and reading themselves are major alternations to the "natural" process of mental development.

Some fairly permanent features do, of course, get laid down in the "critical" period between one and three years of age. For example, if you do not develop binocular vision in those years (because of having crossed eyes or one eye covered), you will not develop it later. However, less extreme limitations, such as language or mathematical deficiencies or music deprivation, are more likely to be amenable to improvement with practice. But to overcome the problems created for us by our axe-made environments, we need to discover when, how, and in what way we can make use of this flexibility to teach old brains new tricks.

Key to this may be the way brain function seems to involve the same kind of serendipitous juxtaposition of data as does the process of innovation itself. Thinking, like innovation, seems to consist of putting things together in new ways but, unlike sequential axemaker-generated logic, imaginative thinking seems to work in a nonlinear way. We do not seem naturally to follow the step-by-step syllogisms of Aristotle or to reduce problems to their smallest parts in the manner of Descartes, or make a list of every possibility like a computer, but instead to flit, haphazard, back and forward across the cortex much as the Irish poet W. B. Yeats described:

> Like the long-legged fly on the stream
> His mind moves upon silence.

As an axemaker might say, there seems to be an "irrational" element to the way the imaginative, nonlogical kind of thinking allows the brain to use less than complete or exact data when it comes to its more intuitive conclusions, or to think in terms of inexact, "fuzzy" intervals such as "almost two," or to make decisions by referring to experience in ways that we may not be able to quantify or even describe. This last is frequently observed when designers of knowledge-based computer systems ask human experts to explain how they do what they do.

Others call this kind of thinking "arational" and see it not as a faculty of mind inferior to logic but one that is complementary to sequential and rational thought. It is the kind of thinking that, on the whole, began to be cut out after the appearance of the first axe. But today, many individuals, from religious thinkers to heads of state or large corporations, are beginning to realize that this capacity to see the world whole, to perceive events as they combine simultaneously, working together with the ability to analyze problems sequentially, may have a vital part to play in securing our future.

Nobody yet knows in detail what "thinking" is, but it seems to involve a process that makes the biggest, fastest computers on Earth look simple. In this regard, it may be worth remembering the saying that the most powerful data processors can only achieve cognitive levels slightly superior to those of a flatworm. Whether the brain operates with a gigantic number of physically interconnected neurons clustered in tiny networks containing "core concepts," or else is structured in hierarchies each dealing with the world by processing it through levels of feature recognition, or with some other system, the number of ways in which the brain's hundred billion neurons can interact may be more than the number of atoms in the universe. And everybody has one of these gigantic systems between their ears.

The reason that the new data-processing systems might bring radical change in our relationship with the axemakers and to the way they have always indirectly organized our society and our thoughts relates to the way the brain works. There would appear to be a match between the non-axemaker, intuitive or "arational" aspects of the thinking processes described above and some of the more complex and interactive capabilities of the next generation of information technology now being developed.

The new systems can present data to the user in the form of a "web" on which all the information contained in a database is interlinked. For example, a simple chain of web data-links might go:

"toilet roll, invented in response to sanitation ceramics, resulting from nineteenth-century sewage developments, triggered by a cholera epidemic, whose social effects generated public health legislation, that established pathology labs, able to function due to tissue-staining techniques, that used aniline dyes, discovered during a search for artificial quinine, in coal-tar that was a by-product of the manufacture of gaslight, that illuminated early workers' evening classes, in factories spinning cotton from America, processed by Eli Whitney's gin, after he developed interchangeable musket parts, that made possible the manufacture of machine tools, for production lines that introduced continuous-process techniques, that one day would make toilet rolls."

Any individual link in this loop of related innovations and events could also provide the start-point for other loops, in which any link could initiate yet other loops and so on.

There are two main attractions to this way of accessing information. First, it is easy to operate because the user can join the web at an entry point matching their level of knowledge and which might therefore might be something as complex as a quantum physics equation or as simple as a toilet roll. Second is the interconnected nature of the web that makes it possible to move from the entry point to anywhere else on the web by a large choice of routes, one of which will best suit the user's own idiosyncratic interests and level of ability.

At each stage of the journey, any link prepares the user for the next link because of the way in which all links relate. Also, at any link there are a number of alternate routes to take, and it is here that the user can make choices based on personal interest or experience. So it is not inconceivable that a journey might begin with the toilet roll and eventually lead to all the data required for understanding quantum physics, or pottery making, or medieval Latin.

Since there would be no "correct" way to arrive at target data designated, say, by curriculum needs, in the kind of educational process that the web might make possible, the web would offer the user a means to "learn" the target information by arriving at it in their own way. "Knowledge" would then be the experience of having traveled on the web, like the knowledge of a city's streets. The journey, therefore, would be more valuable than the destination and relationships between data more valuable than the data. It might be that we would eventually come to value intelligence no longer solely by information-retrieval but by the imaginative way a student constructed such a journey.

The attraction of the web is that the user needs no qualifications to enter, and the process of exploring the web is as easy or complex as the user chooses. The web contains the sum of knowledge, so the experience of a journey links every user in some way to every other user. The number of ways in which a web might be accessed, linked, or restructured could be as many as its users decided.

Use of the web would above all accustom people to become gradually more familiar with the way in which knowledge is not made up of isolated, unconnected "facts," but is part of a dynamic whole. Experience on the web might also bring greater awareness of the social effects of the introduction of any innovation, thanks to the way the result of interrelating data on the web mirrored that of the way innovation affected the community at large. So each time a user journeyed on the web and made new links between data, the new connections would restructure the web in much the same way they might have rearranged society if they had been applied in real terms. In this sense, the web could become a microcosm for society itself. It could serve as a means to play out scenarios for knowledge-manufacture and its potential social effects. Eventually, of course, the web might become the general mode of involvement in all social processes, either in person or through the use of personal electronic "agents." The power of the individual is greatly magnified.

Since the associative nature of a web also reflects, in a limited way, the basic processes of non-axemaker, arational, nonlinear thought, then no initiation would be required to use it. It would offer access to the body of knowledge which is at present available only to those who happen to possess the right "qualifications." Given the linked structure of the web, it would also be unnecessary for users to know exactly what questions to ask. A system of keywords, allied to fuzzy-logic software (the "almost two" approach, as in a typical user's: "I think it has something to do with . . ."), together with prompts, should make it relatively easy to identify the area of potential interest from even the least precise statements or questions. This might be useful, say, when the user was looking for information to help make a political or career decision.

Because of the sheer scale and speed with which information processing of all kinds is growing, massive databases and processing power like those needed to support the web should be available by 2010. Compact discs, each carrying two thousand texts at a cost of less than ten cents per book, are already on the market, and the technology is projected to expand to at least a tenfold increase in storage in the next decade. This means that a major library could be stored on about a hundred discs and made available to students anywhere. Books supplementary to the "basic" two-million library could easily be downloaded from a central source by the technology which today makes video available to people in the developed world.

Because of this, the representation of knowledge first achieved by the axemaker's baton and then expanded through Mesopotamian cuneiform, the alphabet, Gutenberg, Descartes' Method, and television is about to take another quantum leap. And the storage capacity of 2030s nanotechnology will make even the marvelous systems described here look like papyrus.

Many teaching institutions, especially in the United States, are aware of all that the new technology might offer to education and are beginning to acknowledge the limitations of the old reductionist approach to learning, when compared with the potential of these new, relational techniques. In some schools, the first tentative steps are being taken with interdisciplinary first-year courses with names like "Science, Technology and Society."

But use of the web also raises the disturbing issue of what information will be made available to the databases, by whom, and to whom. Much can still be withheld, as corporations are unlikely to allow free entry to their research labs, or politicians willingly to lay all their activities before public scrutiny, or professionals to relinquish the power base of their expertise.

Those in the public media who might make the matter more available for discussion are also subject to pressure from vested interests. Many of the old axemaker tricks have been refined in the twentieth century so as to control an increasingly literate and informed public. A 1920s entry in the *Encyclopaedia of the Social Sciences* proclaimed that "powerful men" should not succumb to "democratic dogmatisms about men being the best judge of their own interests." As recently as Britain's conflict in the Falklands and the U.S. invasion of Panama, Grenada, and Iraq, the blocking of public access to events allowed only selected information to be broadcast. This ensured that the military actions enjoyed initial approval but left the financial and human costs to be accounted for only later when fervor had cooled.

Noam Chomsky, who has studied the control of public information, writes:

> Across a broad spectrum of articulate opinion, the fact that the voice of the people is heard in democratic societies is considered a problem to be overcome by ensuring that the public voice speaks the right words. The general conception is that leaders control us, not that we control them. If the population is out of control and propaganda doesn't work, then the state is forced underground, to clandestine operations and secret wars; the scale of covert operations is often a good measure of popular dissidence.

But Chomsky is in places optimistic: "There are ample opportunities to help create a more humane and decent world, if we choose to act upon them. . . . A democratic communications policy . . . would seek to develop means of expression and interaction that reflect the interests of and concerns of the general population, and to encourage their self-education and their individual and collective action."

A number of reasons offer hope that a democratic communications system such as Chomsky proposes will come about. We have chronicled a few of the many instances in history where institutions have been forced to increase access to information and new ways of thinking by training extra specialists. This process has inevitably (and to some extent, accidentally) also served to increase the level of general education and competence. Also, although specialization has now reached unprecedented levels of complexity, ordinary people are in general more knowledgeable in larger numbers than before, given the amount of access to data they have through newspapers, on-line services, radio, and television. More information-for-rent is coming, including digitized versions of data collections like the Library of Congress, which will give any individual or school access to amounts of information greater than that available to many members of society today. And they'd be able to communicate their feelings over the net.

So with this increasing access to information it should become more difficult than before to propagandize and exclude members of the general public from decision-making processes. And if access to the web were less and less limited, the sheer number of users modifying, adding to, and restructuring the internal linkages of the system might also make it difficult to prevent hackers from breaking any security codes designed to close off access to any part of the web.

One other element might accelerate this process. In order to travel the web, users would not need to "know" anything, in the present sense of being qualified in a single, life-long, specialist expertise. This is not a new turn of events, as it was much discussed (by Plato among others) after the alphabet appeared, when print made memory redundant in the fifteenth century, and when, more recently, calculators became generally available and people relied less on their own ability to calculate. So if the web were able to provide users with the means to find and understand any data in their own terms, then the problem of lay competence in regard to social decision making might be solved.

At present, specialists are trained to advise on everything from Third World agriculture to inner-city job creation to the marketing of new-model cars. In most cases, the process of assessment and then implementation of their suggestions moves top-down. What happens is that those targeted by the decision makers (from scratch-plough farmers buying fertilizer they cannot afford to consumers buying a new machine while the old one still works) may, on occasion, be consulted through polls or even referenda. But too often the surveys are based on the presumption that what most people want is the short-term fix on offer at the time. In any case, the aim of any analysis of public opinion is often little more than to find out how people will react to a product or policy after it goes on the market. The institutional attitude tends to be (to paraphrase Henry Ford): "You can have any color of car as long as it's black."

The process makes perfect sense in an axemaker world, since nobody expects an "unqualified" amateur to contribute to the development of plant genetics or electronic fuel-injection design. But this attitude ignores the fact that in almost every case innovation causes change in the way people live, and in this matter most people are highly competent.

Another important aspect of the issue, besides public involvement in the decision-making process, is the sheer scale of the problems we face. In providing short-term fixes to the problems facing humanity since the axe, perhaps the major distinguishing characteristic of axemaker gifts has been the way they concentrated social control. And since bigger communities meant greater power, leaders and institutions have also been concerned to boost growth. Big has been beautiful since it reached its first apotheosis in Egypt.

However, the limitless-growth concept was born at earlier times when, as we described in the first chapter, the Earth seemed boundless. Back then nobody could have possibly conceived of a "resource limit" that might prevent them from satisfying their desire for growth through expanded food production, bigger transportation systems, larger communication networks, powered by more extensive energy services, to drive faster industrial processes, operated by more and more specialists.

Today the downside of this kind of "progress" is becoming clear as the limitations of the natural world, so long absent from consideration, begin to make themselves felt. The growth ideal has all along ignored the fundamental fact that there is a limit to the extent to which the environment can supply the resources for expansion and absorb its waste products.

Historically, in an axemaker world, with reasonable confidence in economies of scale, authorities have also taken for granted that small social structures are inefficient and have acted against them. As the state grew in power, for instance, it erased independent villages and cities, outlawed local guilds and municipal unions, removed common land, established central coinage and standardized laws, built massive bureaucracies to enforce conformity, and finally provided all the centralized social services that had become necessary because the state had destroyed them at local level. Like all the other gifts that we could not refuse, the growth model became a self-fulfilling prophecy.

The long-term effects of this policy are now damaging the last few remaining self-sustaining communities of the Third World in a way that echoes nineteenth-century Western history. Industrialization is marginalizing local, small-scale village economies, reducing them to the role of city suppliers. And in the urban environment, once-cooperative community members have become competing individuals with no local or family ties, pressured from all sides to conform to Western models and resentful of this.

In the countryside from which these people have come, traditional indigenous knowledge is being dumped. Western-style schooling concentrates on teaching specializations that relate unrealistically to the nature of the local society and its immediate environmental conditions. People are taught Western skills they cannot apply, while losing indigenous skills they need, in classrooms where they are cut off from local and more relevant wisdom accumulated over centuries.

Even those high-profile reformers who advocate "fair-share" redistribution of global wealth maintain a commitment to the continued economic growth and high-technology development, which they consider necessary to generate the wealth needed to pay for their reforms. But the reforms themselves require even greater centralization of control, since the task of redistribution requires supranational authorities. However, as we have tried to show, the end result of such growth-oriented centralizing policies in earlier times has been the environmental and social problems in which the world finds itself today.

One possible solution may lie in the political potential of the web, but this may require a radically different attitude to some long-cherished sacred cows, above all the one that has in recent times maintained axemaker-supported institutions in power by appearing to make that power accountable: representative democracy.

As technology-related social and political issues become ever more complex, if access to relevant information is, as we have suggested, made universal and if electronic agents are available (as they will be within the decade) to act on behalf of the individual, then informed, participatory democracy becomes possible. When the technological means are ready, an educated franchise will demand it, because as Rousseau wrote:

> Sovereignty . . . consists essentially in the general will and the will cannot be represented. Either it is itself, or it is something else; there is no middle ground. The deputies of the people, therefore, are not, nor can they be its representatives; they are merely its agents . . . the instant a people chooses representatives, it is no longer free.

Aristotle thought the *polis* should be small and participation similarly direct:

> If citizens of a state are to judge and distribute office according to merit, then they must know each other's characters; where they do not possess this knowledge . . . the election to office . . . will go wrong.

Information technology is already making this kind of small-scale community more feasible without forcing its members to sacrifice the benefits of modern living. Five eco-villages in Sweden have already separated themselves from the general economic system and are running ecologically based, small-scale alternatives. Other models exist in less radical forms, and the experience of each one of them may have a contribution to make: the worker-owned plywood companies of the Pacific Northwest, the solar-powered community of Davis, California, the consensual democracy of Quaker meetings, the non-hierarchical societies of East Africa, the Amana colonies in Iowa, and others. The web could facilitate this kind of development by giving other communities all the knowledge tools they need to function autonomously.

But the key, political virtue of smaller-scale, web-supported communities run by direct democracy is that they provide forums for debate lost to us since Greece. The debate need not demand personal involvement on the part of every individual. This was the rock on which similar experiments floundered in the 1960s. People then (and most likely people now) were simply too busy to spend time every day with every single item on the community issues. But now techniques are available to create "agents" capable of representing individuals faithfully and comprehensively and of electronically registering (and updating) their views on any and all issues. It may be instructive in this regard to reflect on the long-running axemaker slogan which has so successfully subverted the democratic process. If *vox populi* had ever *really* represented the ultimate authority as it pretended to, why did the phrase end: *vox dei?*

The open social system made possible by small-community participatory democracy is also harder to subvert and control, the structure encourages participation and consensus, and the closer contact between these communities and their environment makes for greater awareness of the need for self-sustaining, non-polluting economies.

All the changes suggested in this chapter so far are complex and wide-ranging, but they do not necessarily require a long time to happen. In general, once any relevant information is available and is understood, people change their lives extraordinarily quickly.

For instance, social scientists in the 1960s wrote papers saying it would take decades of constant government pressure to persuade Americans to alter their reproductive habits. The habit of having as many children as a family could afford was considered fundamental to human nature.

But in the early 1970s, in the United States and Western Europe, the shift to small families only took three years and happened without any need for government pressure. People became aware that economic problems were caused by their reproductive habits, so they changed them.

In terms of the population problem in non-Western cultures, what succeeds may not be free condoms and family planning lectures, neither of which really help Indian peasants to have fewer children, but rather access to information and the social changes that will follow such access: giving women their fair share of power, the likely demand for social security, employment, and the promise of financial independence late in life, which would mean that people need fewer children to look after them in old age.

This is not pie-in-the-sky theory requiring levels of computerized expertise and technical facilities that do not exist outside the advanced nations. It has already happened in the small, low-tech Southwestern Indian state of Kerala and, partly because the government was communist influenced, has been the subject of intense debate for the last two decades.

In the 1970s, development planners dealing with global population problems found in Kerala an exceptional example of a tropical region where both the birth rate and the infant mortality rate were falling, where average life expectancy was approaching seventy years, where the large majority of the adults could read and write, and where women outnumbered men. And all of this had been achieved without violent political change.

Kerala had not industrialized but remained largely dependent on agriculture. Its per capita income was well below the all-India average. The Kerala model suggested that carefully implemented policies stressing equal access to basic needs had improved the inhabitants' lives in the absence of red, green, or industrial revolutions.

In 1975, the U.N. in New York published a report emphasizing "the positive achievements" of Kerala in matters "of equity and of balanced social and economic development." It pointed to the fact that "a relatively poor state" with a low per capita availability of food had made "fairly impressive advances in . . . health and education," which had resulted in "a perceptible difference to the quality of life."

The study made greatest impact on the rest of the world with its evaluation of Kerala's population control success. The report stressed "evidence of a sharp decline in birthrates in Kerala" and suggested this had resulted from "societal changes in attitude to family size resulting from longer life expectation, reduction in infant and child mortality, and female education," which in turn stemmed from substantial government investment in health and education.

As early as the 1920s, a British government analysis had concluded that even the lower castes regarded education as "the door to a new Earth and a new heaven." Kerala's education system stimulated a more demanding, equal-opportunity ethic, which, from the 1930s boosted the spread of education. People wanted education for their children and so governments, increasingly aware of the need to be popular, provided the funds for more schools. For more than a hundred years most Keralans had first encountered political involvement through the school system.

The school system became engrained in Kerala because Kerala's higher castes and sections of local Christians had always believed in the value of education and sent children to schools which the families themselves supported and onto which, after the 1860s, a government-supervised structure was grafted. Also, because lower-caste Hindus associated education with status, they readily attended schools opened by European missionaries. It quickly became obvious that education opened the door to jobs in an expanding commerce and in government service itself.

Decades of widespread public participation in politics stimulated intense political competition after independence in 1947, involving a larger proportion of people than anywhere else in India, as attendance at Kerala's elections strikingly illustrated. The lowest in a state election in Kerala since 1957 was still 8 percent higher than the highest all-India turnout. One observer wrote: "A keen, demanding electorate and frequent elections [for many years they were every two years] have driven Kerala politicians to pursue programs in education, land reform, and health that foster perceptible improvements in people's lives."

Access to information and schooling have been crucial in Kerala. Thousands of women teachers have earned salaries for nearly a century, and the education of girls has raised the age of marriage. Literate women care for their babies more successfully than illiterate mothers. Kerala's schools are its biggest industry, by 1984 commanding 38 percent of the state government's annual expenditure. Schools and colleges are the focus for intense competition between government and non-government structures. These vast educational interests give "communal" organizations an enduring substance and the competition among them has helped to make politics routine and respected.

Writing in 1982 about the Kerala phenomenon, the economist and ex-Ambassador to India, John Kenneth Galbraith, suggested: "It is literacy that comes first. We [economists] had our sequential priorities wrong. We thought we could start with capital investment; we should have started with investment in education."

An Australian writer on Kerala, Robin Jeffrey, concluded in 1992: "Democratic politics, involving large sections of a population, can be made to provide services that people need and, consequently, use. Literate, confident women will, as domestic managers, turn such services into better health for men and women alike. Birth rates will fall. What one yearns for is a further stage in which enough wealth is produced and politics guarantees its fair distribution to ensure far higher measures of well-being for all. Only if Kerala in the 1990s moved visibly in that direction, might it become a model that others would wish to emulate."

Kerala's experience suggests that attitude changes toward population control can be achieved in the Third World without the need for the post-Malthus demographic transition through industrialization that saved the nineteenth-century West. And if humanity's future depends on changing the way people live and the qualities they value, Kerala offers striking evidence that societies can undergo rapid and dramatic transformations and that much chance of any success will likely lie in the hands of emancipated women.

In sum, as we move into the twenty-first century, we face a significant choice: we can continue to rely on the short-term, quick-fix axemaker gifts, or we can begin to think of ourselves as capable individuals who are also part of a global community, because it is on the global scale that humanity is beginning to do damage to the planet. No other species has multiplied and put the environment under stress as has the human animal.

The political fragmentation of the post–Cold War era may be an indication of the direction in which we might proceed. Technology has made the nation state ineffective, unable to protect its citizens militarily from the threat of urban terrorists armed with nuclear knowledge, its economy no longer isolated from the outside world, its laws no longer sovereign, its currency no longer stable.

The fragmentation of these monolithic, centralized communities, where internal cultural anomalies were suppressed for centuries, may present more a promise than a threat. The vociferous ethnic groups emerging from decades of silence reveal a still-flourishing cultural diversity which, like diversity in the natural world, represents a flexibility of response that could be valuable for global survival.

And the more we examine this diversity, the more we are reminded that the cultural spaces between us may not be as wide as we thought. The world's longest-surviving languages are only three thousand years old, religions are almost as young, and the oldest of the so-called "national" identities, that mark out one group from another in the modern world, are relative newcomers. Humanity was in a melting pot for a hundred thousand years before Mesopotamia. A modern Armenian, Mongolian, or Scot only became associated with their "country" fewer than a hundred lifetimes ago. Culturally we are almost entirely Paleolithic, a fact we manifest every time we throw salt over a shoulder, or avoid walking under a ladder, or cross our fingers.

If a late Stone Age baby were to be transported to the present day, dressed appropriately he or she would pass without notice and would be able, in time, to learn all modern skills. So we are closer to the "primitives" in the modern world than we might suspect. What has kept us separate from them has been the overreliance on our sequential axemaker thought-processes and the lack of a common means of communication between us.

In 1703, the German mathematician Gottfried Leibniz wrote in a letter to Reverend Bouvet that he dreamt there would one day be:

> . . . a universal type of writing that would have the advantages of the Chinese system, since each could understand it in his own language, but it would infinitely surpass the Chinese since it would be learned in a few weeks, since its characters are well related according to the order and the connection of things.

From Mesopotamian cuneiform, then through the Greek alphabet, Gutenberg, and especially through the representation of the world in modern computers, we have begun to realize Leibniz's dream. As a result, each person today grows up to master a world that is far more complex and heterogeneous than what might otherwise have been their biological destiny. Since Mesopotamia and Greece, we have learned to represent the world abstractly into strings of letters to be read in one direction. We use mathematics to predict the movement of the stars and space vehicles, as well as to run the computing world we live in. In the process the basic skills of the mind, moving in space, listening and speaking, have become of secondary importance.

And since this is taken for granted, we don't often realize that each child goes through a radical shift in development when they learn to master this artificial world. But it could well be that, given the success of the program Leibniz outlined, mastering the world of information in the twenty-first century will for most people be easier because there will be a new, less-exclusive way to express ourselves. The ability to see relationships and move things around in space may be intellectually as valuable with iconic computers as learning quadratic equations or remembering the table of elements was once with print. When much of the routine drudge-work of the mind is automated, the spatial, intuitive, "navigational" talents may well be much better adapted to accessing knowledge that is structured more like the natural world rather than being reduced to alpha-numeric codes. Reading and writing may well become less important.

For the past few centuries, education has primarily involved the teaching of language and number skills. In part, this is a function of the way information has been displayed, unchanged since the time of the Greek alphabet and Mesopotamian numbers. So learning about Rembrandt or Beethoven or Einstein required the ability to read about them. But if the new systems give individuals more direct access to Rembrandt's painting techniques or to the differing ways symphonies are constructed, then the skills required would be primarily visual, aural, and tactile. With this in mind, it is easier to see the post–Cold War emergence of ethnic pluralism as an opportunity rather than a problem, for with the advent of iconic computers like the Mac, the "literacy" that was to be the hallmark of a learned person, may, for the first time since Greece, no longer be so essential.

Anthropologists have identified a number of characteristics that seem common to most non-technological societies past and present. These societies tend to value practical rather than abstract knowledge, their "primitive" rituals are part of the regular day-to-day realities of life, the groups tend not to support specialists other than the shaman, every member of the group can to some extent do every task, and all share the responsibility for all others. Principally, the "primitive" takes a holist view of life that examines all social decisions for their effect on the community and the environment.

These social values may fit well in the webbed communities of the mid-twenty-first century because they are more appropriate to small, relatively simple social structures that up to now had seemed to be disappearing, as life under the axe became more large-scale and standardized. While we are unlikely to return to Arcadian myths and the bucolic life, the web (and all the support processes it could provide) might make small communities viable once more, functioning in a way that ought to become commonplace all over the planet, where the maxim would be: "Think globally, act locally." And it would take only the kind of currently available renewable energy systems such as solar or geothermal or wind power to make such communities energy-independent and ensure the survival of many cultures that will otherwise face the axe in the next few decades.

As long as the "information superhighways" and the webs to which they would link us are not hijacked by the most powerful-ever information elite in history (and this is a big "if"), they might help return us in some ways to where we were before the axemakers' gifts took the first slice at the world and began applying the process of cut-and-control to both human nature and the nature from which we sprang.

For such communities, the most valuable skills would be generalist rather than specialist. They would prize the ability to connect, to think imaginatively, to understand how data are related, to see patterns in machine-generated innovation, and to assess its social effect before releasing it on society.

These are skills that brains already possess in every community from the brokers of Wall Street to the mud-men of New Guinea. Each human brain has the potential to express many idiosyncratic talents, so in a way each brain is itself an entire culture. Since the time of the first axe the means to make that culture manifest have been suppressed by a relentless pressure to conform that reduced the natural diversity of human self-expression and kept the axemaker mind preeminent.

Today, however, billions of human talents could be on the verge of self-expression if we are willing to take new views and see where they might lead us. And, more than any time before, we have the means to make sure that we all "see"—perceive in a new way. Up to now, when faced with change, we have always been presented with few options. Just as the institutional ability to cut and control was shaped by the axemaker tools of the day, so too was the flexibility of our social systems and the form and extent of our individual response. For instance, only very limited personal freedom of action was possible in Sumeria when the only organizational tool was cuneiform lettering, developed primarily for inventory-keeping. Throughout history, centrally imposed conformity has applied this kind of constraint in order to suppress the expression of anarchic views that could not be adequately managed.

But with the new information technologies, it may be possible easily and swiftly for the community to visualize the pattern of change, to play out the effect of one or other option, and decide which to choose on the basis of more foreknowledge than our ancestors ever had. It might be argued that these scenarios could be as flawed as the suppositions built into them, but nonetheless they are better than nothing, which is what we have had up to now.

Indeed, the approach suggested in this book is only one of a number that might succeed (just as the theme of this book is only one way of looking at the problem). In concentrating on small-scale communities, indigenous knowledge, webbed education, and participatory democracy, for instance, we have not considered other possibilities. There are those who feel that the problems of resource, food, and pollution might be completely solved by genetic engineering, biotechnology, and nanotechnology. Others place their faith in a central world government, population control, and ecological awareness. But whatever the approach, the decision about which problem and solution is most relevant will still have to be made on the basis of the kind of informed and enfranchised community described here.

Whether or not our society takes the course we suggest here will depend on whether or not we can escape the confinement of axemaker thought, and this depends on whether or not we are already too conditioned to be able think arationally as well as rationally.

The first step may be to recognize that we can use our technology as it has been used time and again through history. We can use it to change minds, but this time for our own reasons in our own terms and at our own pace, if we use the coming technologies for what they could be: instruments of freedom. The very interactive nature of the modern world makes it less easy to block such an act and to continue with the old ways of hierarchy and division. But in any case, all that ever kept us in thrall of institutions was our ignorance of the kind of knowledge that could soon now be so easily accessible and understandable that it will be a waste of time to know it. When Gutenberg printed his books, he greatly lessened the power of memory and tradition. The new technologies will lessen the power of arcane, specialist knowledge. And when they do, we will all, in one sense, return to what we were before the first axe.

The culture we live in, based on the sequential influence of language on thought and operating according to the rationalist rules of Greek philosophy and reductionist practice, has wielded tremendous power. It has given us the wonders of the modern world on a plate. But it has also fostered beliefs that have tied us to centralized institutions and powerful individuals for centuries, which we must shuck off if we are to adapt to the world we've made: that unabated extraction of planetary resources is possible, that the most valuable members of society are specialists, that people cannot survive without leaders, that the body is mechanistic and can only be healed with knives and drugs, that there is only one superior truth, that the only important human abilities lie in the sequential and analytic mode of thought, and that the mind works like an axemaker's gift.

Above all (and most recently) we have also been persuaded to think that it is unacceptable to be different or even to acknowledge that differences in abilities exist between us. But our survival may depend on the realization and expression of humanity's immense diversity. Only if we use what may be the ultimate of the many axemaker's gifts—the coming information systems—to nurture this individual and cultural diversity, only if we celebrate our differences rather than suppressing them, will we stand a chance of harnessing the wealth of human talent that has been ignored for millennia and that is now eager, all around the world, for release.

Select Bibliography

Prologue

Idries Shah. *Reflections.* 2d ed. London: The Octagon Press, 1969.

Chapter 1: Getting an Edge

Armelagos, G. J., et al. "Effects of Nutritional Change on Skeletal Biology." In *Hunters to Farmers,* edited by Brandt, S. A., and J. D. Clark. Berkeley: University of California Press, 1984.
Berlin, B., and P. Kay. *Basic Color Terms: Their Universality and Evolution.* Berkeley: University of California Press, 1969.
Brace, C. L., and R. J. Hinton. "Oceanic Tooth Size Variation as a Reflection of Biological and Cultural Mixing." *Current Anthropology* 22(5).
Brandt, S. A., and J. D. Clark, eds. *Hunters to Farmers.* Berkeley: University of California Press, 1984.
Cavalli-Sforza, C. L., and M. W. Feldman. *Cultural Transmission and Evolution.* Princeton: Princeton University Press, 1981.
Cohen, J. *The Privileged Ape.* Lancashire: Parthenon Publications, 1989.
Cosmides, L., and J. Tooby. "From Evolution to Behavior: Evolutionary Psychology as the Missing Link." In *The Latest on the Best: Essays on Evolution and Optimality,* edited by J. Dupre. Cambridge: MIT Press, 1986.
DeLumley, H. "A Paleolithic Camp at Nice." *Scientific American* 220(5) (1969).
Desmond, K. *The Harwin Chronology of Inventions, Innovations, Discoveries.* London: Constable, 1986.
Donald, M. *Origins of the Modern Mind.* Cambridge: Harvard University Press, 1991.
Foley, R., ed. *Hominid Evolution and Community Ecology.* London: Academic Press, 1984.
Foster, M. Lecron. "Symbolic Origin and Transition in the Paleolithic." In *Origin of Modern Humans,* edited by P. Mellars. Edinburgh: Edinburgh University Press, 1990.
Goody, J. *Production and Reproduction.* Cambridge: Cambridge University Press, 1976.
Gowlett, J. G. *Ascent to Civilisation.* London: Collins, 1984.
Haldane, J. B. S. *The Causes of Evolution.* London: Longmans Green, 1932.

Hausman, A. J. "Holocene Human Evolution in Africa." In *Hunters to Farmers,* edited by Brandt, S. A., and J. D. Clark. Berkeley: University of California Press, 1984.
Hopkins, D. M., ed. *The Bering Land Bridge.* Stanford: Stanford University Press, 1967.
Hubel, D. H., and T. N. Wiesel. "Receptive Fields, Binocular Interactions and Functional Architecture in the Cat's Visual Cortex." *Journal of Physiology* 160 (1962).
Johnson, A. "In Search of the Affluent Society." *Human Nature* 1(9), (1978).
Lancaster, J. L. "Carrying and Sharing in Human Evolution." *Human Nature* 1(2), (1978).
Lean, G., and D. Hinrichsen. *Atlas of the Environment.* Oxford: WWF/Helicon, 1992.
Lovejoy, C. O. "The Gait of Australopithecines." *Yearbook of Physical Anthropology* 17 (1974).
———. "The Origin of Man." *Science* 211 (1981).
Marshak, A. "The Art and Symbols of Ice Age Man." *Human Nature* 1(9), (1978).
Mellen, S. L. W. *The Evolution of Love.* San Francisco: W. H. Freeman, 1981.
Renfrew, C. "World Languages and Human Dispersals—a Minimalist View." In *Transitions to Modernity,* edited by Hall, J., and I. C. Jarvie. Cambridge: Cambridge University Press, 1992.
Sahlins, M. *Stone Age Economics.* Chicago: Aldine, 1972.
Stocking, G. W. *Race, Culture and Evolution.* Chicago: University of Chicago Press, 1982.
Stringer, C., and C. Gamble. *In Search of the Neanderthals.* London: Thames & Hudson, 1993.
Terrell, J. *Prehistory in the Pacific Islands.* Cambridge: Cambridge University Press, 1986.
Turnbull, C. "Some Observations Regarding the Experiences and Behavior of the Bambuti Pygmies." *American Journal of Psychology* 74 (1961).
Underwood, J. H. *Human Variation and Human Microevolution.* New York: Prentice Hall, 1979.
Washburn, S. "Tools and Human Evolution." *Scientific American* 203(3), (1960).
White, T. D., et al. "*Australopithecus ramidus,* a New Species of Early Hominid from Aramis, Ethiopia." *Nature* (September 1994).
Wilson, E. O. *Sociobiology.* Cambridge: Harvard University Press, 1975.

CHAPTER 2: TOKEN CONTRIBUTION

Adams, Barbara. *Predynastic Egypt.* Princes Risborough: Shire, 1988.
Alizadeh, Abbas. "Socio-Economic Complexity in Southwestern Iran During the Fifth and Fourth Millennia B.C.: The Evidence from Tall-i Bakun A." *Iran* 24 (1988).
Astle, Thomas. *The Origin and Progress of Writing.* 2d ed. London: J. White, 1803.
Cherry, Colin. *On Human Communication.* London: John Wiley and Sons, 1957.
Chiera, Edward. *They Wrote on Clay.* Chicago: University of Chicago Press, 1938.
Childe, G. *Man Makes Himself.* Harmondsworth: Penguin, 1936.
Claessen, Henri J. M., and Peter Skalnik. *The Early State.* The Hague: Mouton, 1978.
Clark, Grahame. *World Prehistory: A New Outline.* Cambridge: Cambridge University Press, 1969.
Diakonoff, Igor M. "Some Reflections on Numerals in Sumerian Towards a History of Mathematical Speculations." *Journal of the American Oriental Society* 103, no. 1, (1983).
Driver, G. R. *Assyrian Laws.* Oxford: Clarendon Press, 1935.

Driver, G. R., and J. C. Miles. *The Babylonian Laws.* Vols 1 & 11. Oxford: Clarendon Press, 1952.
Finlay, F. *Ancient Slavery and Modern Ideology.* Harmondsworth: Penguin, 1983.
Frankfort, Henri. *The Birth of Civilization in the Near East.* London: Williams and Norgate, 1951.
Fry, Edmund. *Pantagraphia.* London: Cooper and Wilson, 1977.
Haas, Jonathan. *The Evolution of the Prehistoric State.* New York: Columbia University Press, 1982.
Henry, Donald O. *From Foraging to Agriculture.* Philadelphia: University of Pennsylvania Press, 1989.
Innis, Harold A. *Empire and Communications.* Oxford: Clarendon Press, 1950.
Jackson, Donald. *The Story of Writing.* New York: Taplinger Publishing Company, 1981.
Jasim, Sabah Aboud, and Joan Oates. "Early Tokens and Tablets in Mesopotamia: New Information from Tell Abada and Tell Brak." *World Archaeology* 17, no. 3, (1986).
Kramer, S. N. *The Sumerians.* Chicago: University of Chicago Press, 1963.
Kraus, R. C. *Brushes with Power.* Berkeley: University of California Press, 1991.
Kutas, M., S. A. Hillyard, and M. S. Gazzaniga. "Processing of Semantic Anomaly by Right and Left Hemispheres of Commissurotomy Patients: Evidence from Event-related Brain Potentials." *Brain* 111 (June 1988).
Lieberman, S. J. "Of Clay Pebbles, Hollow Clay Balls, and Writing: A Sumerian View." *American Journal of Archaeology* 84, no. 3, (1980).
MacDowel, Douglas M. *The Law in Classical Athens.* London: Thames & Hudson, 1978.
Macneill, William. *The Rise of the West.* Chicago: University of Chicago Press, 1963.
Maisels, Charles Keith. *The Emergence of Civilization: From Hunting and Gathering to Agriculture, Cities, and the State in the Near East.* London: Routledge & Kegan Paul, 1990.
Menninger, Karl. *Number Words and Number Symbols.* Cambridge: MIT Press, 1977.
Mote, Frederick W. *The Intellectual Foundations of China.* New York: Alfred A. Knopf, 1989.
Naquin, Susan, and E. S. Rawski. *Chinese Society in the Eighteenth Century.* New Haven: Yale University Press, 1987.
Neufeld, E. *The Hittite Laws.* London: Luzac & Co., 1951.
Pfeiffer, J. E. *The Emergence of Society: a Prehistory of the Establishment.* New York: Harper & Row, 1977.
Robson, William A. *Civilisation and the Growth of Law.* London: Macmillan, 1935.
Sampson, Geoffrey. *Writing Systems.* Stanford: Stanford University Press, 1985.
Schmandt-Besserat, Denise. "From Tokens to Tablets: A Re-evaluation of the So-called Numerical Tablets." *Visible Language* 15, no. 4, (1981).
———. "The Emergence of Recording." *American Anthropologist* 84, no. 4, (1982).
———. "The Envelopes that Bear the First Writing." *Technology and Culture* 21, no. 3, (1980).
———. "Tokens at Uruk." Cat. nos. 23, 104, and 107; Susa Cat. nos. 170, 658, and 659, BWII, (1988).
———. "The Use of Clay Before Pottery in the Zagros." *Expedition* 16, no. 2, (1974).
———. *Before Writing:* Volume 1, *From Counting to Cuneiform.* Austin: University of Texas Press, 1992.
Service, Elman R. *Origins of the State and Civilization: The Process of Cultural Evolution.* New York: Norton, 1975.
Slobin, D. I. "Universals of Grammatical Development in Children." *Advances in Psycho-*

linguistics, edited by Flores d'Arcais, G. B., and W. J. M. Levelt. Amsterdam: North-Holland Publishing, 1970.

Smith, Philip E. L. "Ganj Dereh Tepe." *Paléorient* 2, no. 1, (1974).

Solecki, Ralph S. *Shanidar.* Harmondsworth: Penguin, 1972.

Trigger, B. G., et al. *Ancient Eygpt: A Social History.* Cambridge: Cambridge University Press, 1983.

Vallat, François R. *Le Natoufien.* Paris: Cahiers de la Revue Biblique, no. 15 (1975).

Wright, Henry T. "An Early Town on the Deh Luran Plain, Excavations at Tepe Farukhabad." *Memoirs of the Museum of Anthropology* 13. Ann Arbor: University of Michigan, 1981.

CHAPTER 3: THE ABC OF LOGIC

Baron, J., and C. Strawson. "Use of Orthographic and Word-specific Knowledge in Reading Words Aloud." *Journal of Experimental Psychology: Human Perceptual Performance* 2 (1976).

Carmon, A. "Temporal Processing and the Left Hemisphere." *The Behavioral and Brain Sciences* 4 (1981).

Cleve, F. M. *The Giants of Pre-Sophistic Greek Philosophy.* 2 vols. The Hague: Martinus Nijhoff, 1965.

Cline, H. E. *Introducing Logic, Epistemology and Ethics.* Lanham: University Press of America, 1987.

Cook S., and F. Adcock. *The Cambridge Ancient History.* Vol. 12. Cambridge: Cambridge University Press, 1981.

Cook, S., F. Adcock., and M. Charlesworth, eds. *The Cambridge Ancient History.* Vol. 10 Cambridge: Cambridge University Press, 1985.

Cornford, F. M. *From Religion to Philosophy.* Hassocks: Harvester Press, 1980.

Daniels, P. "A Calligraphic Approach to Aramaic Paleography." *Journal of Near Eastern Studies* 43 (1984).

deKerckhove, D. "A Theory of Greek Tragedy." *Sub-Stance* 29 (1981).

———. "Effets cognitifs de l'alphabet." In *Pour Comprendre,* edited by D. deKreckhove, and D. Jutras. Ottawa: UNESCO, 1984.

deKreckhove, D., and C. J. Lumsden, eds. *The Alphabet and the Brain.* Heidelberg: Springer-Verlag, 1988.

Edelman, G. M., and L. H. Finkel. "Neuronal Group Selection in the Cerebral Cortex." In *Dynamic Aspects of Neocortical Function,* edited by G. M. Edelman, W. E. Gall, and W. M. Cowan. New York: Wiley, 1984.

Endo, M., I. Shimizu, and I. Nakamura. "Laterality Differences in Recognition of Japanese and Hangul Words by Monolinguals and Bilinguals." *Cortex* 17 (1981).

Farrington, B. *Aristotle, Founder of Scientific Philosophy.* London: Weidenfeld & Nicholson, 1965.

———. *Science in Antiquity.* Oxford: Oxford University Press, 1969.

Flaceliere, R. *Daily Life in Greece in the Time of Pericles.* London: Weidenfeld & Nicholson, 1965.

Gainotti, G., et al. "Selective Semantic-lexical Impairment of Language Comprehension in Right-brain Damaged Patients." *Brain and Language* 13 (1981).

Galin, D., and R. Ornstein. "Lateral Specialization of Cognitive Mode: An EEG Study." *Psychophysiology* 9 (1972).

Goody, J., and I. Watt. "The Consequences of Literacy." *Comparative Studies in Society and History* 5 (1963).

Graff, H. J. "The History of Literacy: Towards the Third Generation." *Interchange* 17 (1986).

Gregory, J. C. *A Short History of Atomism.* London: A. & C. Black, 1931.

Guthrie, W. K. C. *The Greek Philosophers.* London: Methuen, 1950.

Havelock, E. A. *The Muse Learns to Write. Reflections on Orality and Literacy from Antiquity to the Present.* New Haven & London: Yale University Press, 1986.

———. "The Alphabetic Mind: a Gift of Greece to the Modern World." *Oral Tradition* 1 (1986).

———. *The Literate Revolution in Greece and its Cultural Consequences.* Princeton: Princeton University Press, 1982.

Hayashi, R., and T. Hatta. "Visual Field Differences in a Deeper Semantic Processing Task with Kanji Stimuli." *Japanese Psychological Research* 24 (1982).

Hippocrate, G., trans. *Aristotle's Metaphysics.* Bloomington & London: Apostle, 1966.

Lafont, R., et al. *Anthropologie de l'écriture.* Paris: Centre George Pompidou, 1984.

Lesser, G. S. "Cultural Differences in Learning and Thinking Styles." In *Individuality in Learning and Thinking,* edited by S. Messick. San Francisco: Jossey-Bass, 1976.

Lloyd, G. E. R. *Early Greek Science.* London: Chatto & Windus, 1970.

McCarter, F. K. "The Early Diffusion of the Alphabet." *Biblical Archaeologist* 37 (1974).

Millard, A. R. "The Canaanite Linear Alphabet and Its Passage to the Greeks." *Kadmos* 15 (1976).

Naveh, J. "Some Semitic Epigraphical Considerations on the Antiquity of the Greek Alphabet." *American Journal of Archaeology* 77 (1973).

———. *Early History of the Alphabet: An Introduction to West Semitic Epigraphy and Palaeography.* Jersualem: Magnes, 1982.

Nebes, R. "Dominance of the Minor Hemisphere in Commissurotomized Man in a Test of Figural Unification." *Brain* 95 (1972).

Olson, D. R. "The Cognitive Consequences of Literacy." *Canadian Psychology/Psychologie Canadienne* 27 (1986).

Ong, W. J. *Rhetoric, Romance and Technology: Studies in the Interaction of Expression and Culture.* Ithaca: Cornell University Press, 1971.

———. *The Presence of the Word.* New Haven: Yale University Press, 1967.

Russell, B. *History of Western Philosophy.* Woking: Unwin, 1946.

Sasanuma, S. "Kana and Kanji Processing in Japanese Aphasics." *Brain and Language* 2 (1975).

Scribner, S., and M. Cole. *The Psychology of Literacy.* Cambridge: Harvard University Press, 1981.

Skoyles, J. R. "Alphabet and the Western Mind." *Nature* 309 (1984).

Sperry, R. W. "Neurology and the Mind-Brain Problem." *American Scientist* 40 (1952).

———. "Some Effects of Disconnecting the Cerebral Hemispheres." *Science* 217 (1982).

Taylor, I. "Psychology of Literacy: East and West." In *The Alphabet and the Brain,* edited by deKerckhove, D., and C. J. Lumsden. Heidelberg: Springer-Verlag, 1988.

Tsunoda, T. *The Japanese Brain: Working Mechanisms of Brain and East-West Culture.* Tokyo: Taishu Kan, 1978.

Whorf, B. *Language, Thought, and Reality: Selected Writings of Benjamin Lee Whorf,* edited by J. B. Carroll. New York: Wiley, 1942.

Witelson, S. F. "Sex and the Single Hemisphere: Specialization of the Right Hemisphere for Spatial Processing." *Science* 193 (1976).

Woodbury, L. "The Literate Revolution: A Review Article." *Classical Views/Echos du Monde Classique* 27 (1983).

CHAPTER 4: FAITH OF POWER

Atroshenko, V. I., and J. Collins. *The Origins of the Romanesque.* London: Lund Humphries, 1981.
Barmby, Rev. James, ed. and trans. *A Select Library of Nicene and Post-Nicene Fathers of the Christian Church.* Vol. 13. Oxford: Oxford University Press, 1898.
Birks, P. *Justinian's Institutes.* London: Duckworth, 1987.
Brunt, P. *Roman Imperial Themes.* Oxford: Clarendon, 1990.
Bruun, C. *The Water Supply of Ancient Rome.* Helsinki: Soc. Scientia Fenica, 1991.
Burtt, E. A. *The Metaphysical Foundations of Modern Science.* London: Routledge & Kegan Paul, 1972.
Cambridge Ancient History. Vol. 10. "The Augustan Empire 44 BC–AD 70," edited by Cook, S., F. Adcock, and M. Charlesworth. Cambridge: Cambridge University Press, 1985.
Cambridge Ancient History. Vol XII. "The Imperial Crisis and Recovery AD 193–323," edited by Adcock, F., and S. Cook. Cambridge: Cambridge University Press, 1981.
Cameron, Averil. *The Later Roman Empire.* London: Fontana, 1993.
Chenu, M. D. *Nature, Man and Society in the Twelfth Century.* Chicago: University of Chicago Press, 1957.
Diehl, C. *Byzantium, Greatness and Decline.* New Brunswick: Rutgers University Press, 1957.
Dudley, R. *Roman Society.* Harmondsworth: Penguin, 1975.
Eales, Samuel J., ed. and trans. *The Life and Works of St Bernard, Abbot of Clairvaux.* 4 vols. London, 1889.
Finley, M. *Politics in the Ancient World.* Cambridge: Cambridge University Press, 1983.
Geanakoplos, D. J. *Constantinople and the West.* Madison: University of Wisconsin Press, 1989.
———. *Interaction of the "Sibling" Byzantine and Western Cultures in the Middle Ages 330–1600.* New Haven: Yale University Press, 1976.
Grabar, A. *L'Empereur dans l'art byzantin.* London: Variorum, 1971.
Grant, E. *Studies in Medieval Science and Natural Philosophy.* London: Variorum, 1981.
Haskins, C. *The Renaissance of the Twelfth Century.* Cambridge: Harvard University Press, 1927.
Hockey, S. F., ed. *The Cartulary of Carisbrooke Priory.* Isle of Wight: County Records Office, 1981.
Holland, J., and J. Monroe. *The Order of Rome.* London: Cassell, 1980.
Jansen, Sue Curry. *Censorship: the Knot that Binds Power and Knowledge.* Oxford: Oxford University Press, 1991.
Jones, A. H. *The Later Roman Empire 284–602.* 2 vols. Oxford: Blackwell, 1973.
Kempers, B. *Kempers Painting Power and Patronage.* London: The Penguin Press, 1992.
Le Goff, Jacques. *Medieval Civilization.* Oxford: Blackwell, 1988.
Leighton, Albert C. *Transport and Communication in Early Medieval Europe AD 500–1100.* Newton Abbot: David & Charles, 1972.
Lindberg, David C. *The Beginnings of Western Science.* Chicago: University of Chicago Press, 1992.

Lynch, Joseph H. *The Medieval Church: A Brief History.* London and New York: Longman, 1992.

MacCabe, Joseph. *The History and Meaning of the Catholic Index of Forbidden Books.* Girard, KA: Haldeman-Julius Pubs., 1931.

Madski, G. *Preaching and Propaganda in the Middle Ages.* Paris: Presses Universitaires de France, 1983.

Manchester, William. *A World Lit Only By Fire: the Medieval Mind and the Renaissance.* Boston: Little, Brown, 1992.

Mango, C. *Byzantium and Its Image.* London: Variorum, 1984.

Mann, Michael. *The Sources of Social Power. Vol 1, A History of Power to AD 1760.* Cambridge: Cambridge University Press, 1986.

Manschreck, Clyde L. *A History of Christianity in the World.* Englewood Cliffs and London: Prentice Hall, 1985.

Marrou, H. I. *A History of Education in Antiquity.* London and New York: Sheed and Ward, 1956.

Meyendorff, J. *Byzantium and the Rise of Russia.* Cambridge: Cambridge University Press, 1981.

Milis, Ludo J. R. *Angelic Monks and Earthly Men: Monasticism and Its Meaning in Medieval Society.* Woodbridge: Boydell Press, 1992.

Morris, Colin. *The Discovery of the Individual, 1050–1200.* New York and Burlingame: Harcourt, Brace & World, 1972.

Mullett, M. and R. Scott. *Byzantium and the Classical Tradition.* Birmingham: Centre for Birmingham Studies, 1981.

Nicol, D. M. *Byzantium and Venice.* Cambridge: Cambridge University Press, 1988.

Noble, Thomas F. X. "Literacy and the Papal Government in Late Antiquity and the Early Middle Ages." In *The Uses of Literacy in Early Mediaeval Europe,* edited by Rosamund McKitterick. Cambridge: Cambridge University Press, 1990.

Pattison, Robert. *On Literacy: The Politics of the Word from Homer to the Age of Rock.* Oxford: Oxford University Press, 1982.

Reynolds, L. D., and N. G. Wilson. *Scribes and Scholars. A Guide to the Transmission of Greek and Latin Literature.* London: Oxford University Press, 1968.

Rich, Pierre. *Education and Culture in the Barbarian West.* Columbia: University of South Carolina Press, 1976.

Richards, Jeffrey. *Consul of God: The Life and Times of Gregory the Great.* London: Routledge & Kegan Paul, 1980.

Setton, K. M. "The Byzantine Background to the Italian Renaissance." *Proc. Amer. Philosophical Society,* (1956).

Sinnigen, W. G., and A. E. Boak. *History of Rome to AD 565.* New York: Macmillan, 1977.

Smart, A. *The Dawn of Italian Painting.* London: Phaidon, 1978.

Sorrell, Roger D. *St Francis of Assisi and Nature.* Oxford: Oxford University Press, 1988.

Southern, R. W. *Western Society and the Church in the Middle Ages.* Harmondsworth: Penguin, 1970.

Stock, Brian. *The Implications of Literacy: Written Language and Models of Interpretation in the Eleventh and Twelfth Centuries.* Princeton: Princeton University Press, 1983.

Unger, R. *Dutch Shipbuilding Before 1800.* Assen: Van Gorcum, 1978.

Chapter 5: Fit to Print

Arnove, R. F., and H. J. Graff, eds. *National Literacy Campaigns.* New York: Plenum, 1987.

Beardsely, T. G. "The Classics and Their Spanish Translation." *Renaissance and Reformation* 8(1), (1971).

Bloch, Maurice. *Feudal Society.* London: Routledge & Kegan Paul, 1960.

Bowen, James. *A History of Western Education.* Vols. 1–3. London: Methuen, 1972–1981.

Boxer, C. R. *The Portuguese Seaborne Empire.* London: Hutchinson, 1977.

Breuilly, John. *Nationalism and the State.* 2d ed. Manchester: Manchester University Press, 1993.

Campbell, T. *The Earliest Printed Maps 1472–1500.* London: British Library Publications, 1987.

Carter, H. *A View of Early Typography to 1600.* Oxford: Clarendon, 1969.

Chartier, R. *The Cultural Uses of Print in Early Modern France.* Princeton: Princeton University Press, 1987.

Chrisman, M. V. *Lay Culture, Learned Culture 1480–1599.* New Haven: Yale University Press, 1982.

Cipolla, C. M. *Literacy and Development in the West.* Harmondsworth: Penguin, 1969.

Cohn, Henry J., ed. *Government in Reformation Europe, 1520–1560.* London: Macmillan, 1971.

Coleman, D. *The British Paper Industry 1495–1860.* Oxford: Clarendon, 1958.

Coulton, G. G. *Art and the Reformation.* Cambridge: Cambridge University Press, 1953.

Cressy, D. *Literacy and Social Order.* Cambridge: Cambridge University Press, 1980.

Davis, N. Z., ed. "Printing and the People." In *Society and Culture in Early Modern France.* London: Duckworth, 1975.

DuBellay, Joachim. *Defence and Illustration of the French Language.* Paris, 1549.

Eisenstein, Elizabeth L. *The Printing Press as an Agent of Change.* Vols 1 and 2. Cambridge: Cambridge University Press, 1979.

George, K. and C. George. *The Protestant Mind of the English Reformation 1570–1640.* Princeton: Princeton University Press, 1961.

Gillespie, C. C. *The Edge of Objectivity.* Princeton: Princeton University Press, 1960.

Graff, H. J. *The Legacies of Literacy.* Bloomington: Indiana University Press, 1987.

Greengrass, Mark, ed. *Conquest and Coalescence: The Shaping of the State in Early Modern Europe.* London: Edward Arnold, 1991.

Hill, C. *Society and Puritanism in Prerevolutionary England.* Harmondsworth: Penguin, 1986.

Hirsch, R. *Printing, Selling and Reading 1450–1500.* Wiesbaden: Otto Harrassowitz, 1967.

Hortman, François. *Franco-Gallia.* Paris, 1573.

Houston, R. A. *Literacy in Early Modern Europe. Culture and Education, 1500–1800.* London: Longman, 1988.

Hume, Robert. *Education Since 1700.* Oxford: Heinemann Educational, 1989.

Kernan, A. *Samuel Johnson and the Impact of Print.* Princeton: Princeton University Press, 1987.

Klaits, Joseph. *Printed Propaganda Under Louis XIV.* Princeton: Princeton University Press, 1976.

Levine, K. *The Social Context of Literacy.* London: Routledge & Kegan Paul, 1986.

Lister, R. *Old Maps and Globes 1500–1850.* London: Bell and Hyman, 1977.

Loades, David. *Politics, Censorship and the English Reformation.* London: Pinter, 1991.
Lowry, M. *The World of Aldus Manutius.* Oxford: Blackwell, 1979.
Luke, C. *Pedagogy, Printing and Protestantism.* Albany: SUNY Press, 1989.
McLuhan, Marshall. *The Gutenberg Galaxy.* London: Routledge & Kegan Paul, 1967.
Mann, Michael. *The Sources of Social Power.* Vols 1 and 2. Cambridge: Cambridge University Press, 1986.
Marker, G. *Publishing, Printing and the Origins of Intellectual Life in Russia 1700–1800.* Princeton: Princeton University Press, 1985.
Martin, H. J., and L. Febvre. *The Coming of the Book 1450–1800.* Translated by Gerard, D. London: NLB, 1976.
Neuberg, Victor E. *Popular Education in 18th Century England.* London: Woburn Press, 1971.
Orme, Nicholas. *From Childhood to Chivalry.* London: Methuen, 1984.
Perkinson, Henry J. *Since Socrates: Studies in the History of Western Educational Thought.* New York & London: Longman, 1980.
Porter, Roy, and Mikuls Teich, eds. *The Renaissance in National Context.* Cambridge: Cambridge University Press, 1992.
Prest, W. *The Professions in Early Modern England.* London: Croom Helm, 1987.
Price, D. de Solla. "The Book as a Scientific Instrument." *Science* 158 (October 1967).
Rice, Eugene, F., Jr. *The Foundations of Early Modern Europe, 1460–1559.* London: Weidenfeld & Nicholson, 1970.
Stone, Lawrence. "Literacy and Education in England, 1640–1900." *Past and Present* 42.
Thomas, Keith. *Rule and Misrule in the Schools of Early Modern England.* Reading: University of Reading, 1975.
Thompson, Francis. *Saint Ignatius Loyola.* London: Burns & Oates, 1909.
Townsend, G., ed. *Acts and Monuments of John Foxe.* London, 1885.
Trevor Davies, R. *The Golden Century of Spain, 1501–1621.* London: Macmillan, 1937.
Tyson, G., and S. Wagonheim. *Print and Culture in the Renaissance.* Cranberry, NJ: University of Delaware Press, 1986.
Wall, John N. "The Reformation in England and the Typographical Revolution." In *Print and Culture in the Renaissance,* edited by Tyson, G. P., and S. Wagonheim. Cranberry, NJ: University of Delaware Press, 1986.
Woodward, W. H. *Desiderius Erasmus: Concerning the Aim and Method of Education.* Cambridge: Cambridge University Press, 1904.
Wright, A. D. *The Counter-Reformation.* London: Weidenfeld & Nicholson, 1982.

CHAPTER 6: NEW WORLDS

Barnes, B., and S. Shapin. *Natural Order.* Beverly Hills and London: Sage, 1979.
Bordo, S. R. *The Flight to Objectivity.* Albany: SUNY Press, 1989.
Burtt, E. A. *The Metaphysical Foundations of Modern Science.* London: Routledge & Kegan Paul, 1972.
Clark, G. N. *The Seventeenth Century.* Oxford: Clarendon Press, 1947.
Cole, Charles W. *Colbert and a Century of French Mercantilism.* New York: Columbia University Press, 1939.
Cromwell, Oliver. *Letters and Speeches,* edited by Lomas, S. C. 2 vols. London: Methuen, 1904.
Da Silva, C. R. *The Portuguese in Ceylon.* Colombo: H. W. Cane & Co., 1972.

Davis, J. C. *Utopia and the Ideal Society: A Study of English Utopian Writing, 1516–1700.* Cambridge: Cambridge University Press, 1981.
Easlea, B. *Witch-hunting, Magic and the New Philosophy.* Brighton: Harvester Press, 1980.
Eliav-Feldon, Miriam. *Realistic Utopias: The Ideal Imaginary Societies of the Renaissance 1516–1630.* Oxford: Oxford University Press, 1982.
Erasmus, Charles J. *In Search of the Common Good: Utopian Experiments Past and Future.* New York: Free Press, 1977.
Figlio, K. "Theories of Perception and Physiology of the Mind in the Late Eighteenth Century." *History of Science* 12 (1975).
Frangsmyr, T. *Linnaeus, the Man and His Work.* Berkeley: University of California Press, 1983.
Garber, D. "Science and Certainty in Descartes." In *Descartes, Critical and Interpretive Essays,* edited by M. Hooker. Baltimore: Johns Hopkins University Press, 1980.
Gillespie, C. C. *The Edge of Objectivity.* Princeton: Princeton University Press, 1960.
Hazard, P. *The European Mind 1680–1715.* Harmondsworth: Penguin, 1973.
Hecksher, Eli. *Mercantilism.* London: Allen & Unwin, 1935.
Hill, C. *Intellectual Origins of the English Revolution.* London: Panther, 1972.
Hooker, M. *Descartes. Critical Essays.* Baltimore: Johns Hopkins University Press, 1978.
Jacob, Margaret C. *The Newtonians and the English Revolution.* Brighton: Harvester Press, 1976.
Knight, D. *Ordering the World.* London: Burnett Books, 1981.
Lindberg, D. C., and R. S. Westman, eds. *Reappraisals of the Scientific Revolution.* Cambridge: Cambridge University Press, 1990.
Manuel, F. E., and F. P. *Utopian Thought in the Western World.* Oxford: Blackwell, 1979.
Merchant, C. *The Death of Nature.* London: Harper and Row, 1980.
Moravia, S. "The Enlightenment and the Sciences of Man." *History of Science* 18 (1980).
Moraze, C. *The Nineteenth Century.* London: Allen & Unwin, 1976.
Morrill, John, ed. *Oliver Cromwell and the English Revolution.* Harlow: Longman, 1990.
Perkinson, Henry J. *Since Socrates: Studies in the History of Western Educational Thought.* New York & London: Longman, 1980.
Schuster, J. A., and R. R. Yeo. *The Politics and Rhetoric of Scientific Method.* Dordrecht & Lancaster: Reidel, 1986.
Seligman, R., and A. Johnson. *Dictionary of Social Sciences.* London: Macmillan, 1934.
Shafer, Boyd C. *Faces of Nationalism: New Realities and Old Myths.* New York: Harcourt, Brace, Jovanovich, 1972.
Shapin, Steven, and S. Schaffer. *Leviathan and the Air Pump.* Princeton: Princeton University Press, 1985.
Shapiro, B. J. *Probability and Certainty in Seventeenth Century England.* Princeton: Princeton University Press, 1983.
Slaughter, M. *Universal Languages and Scientific Taxonomy in the Seventeenth Century.* Cambridge: Cambridge University Press, 1982.
Smith, David L. *Oliver Cromwell: Politics and Religion in the English Revolution, 1640–1658.* Cambridge: Cambridge University Press, 1991.
Surtz, Edward, ed. *Utopia.* New Haven: Yale University Press, 1967.
Sydenham, P. H. *Measuring Instruments: Tools of Knowledge and Control.* London: Peter Peregrinus, 1979.
Treitschke. P., *History of Germany in the Nineteenth Century.* Translated by Paul Eden, and Cedar Paul. London: Jarrold & Sons, 1915.
Webster, C. *The Great Instauration.* London: Duckworth, 1975.

Westfall, R. S. *Construction of Modern Science.* Cambridge: Cambridge University Press, 1977.
Whitney, C. *Francis Bacon and Modernity.* New Haven: Yale University Press, 1986.
Williams, B. *Descartes, the Project of Pure Enquiry.* Brighton: Harvester, 1978.

Chapter 7: Root and Branch

Allan, D. G. C. *William Shipley, Founder of the Royal Society of Arts.* London: Scholar Press, 1979.
Barbour, V. *Capitalism in Amsterdam in the 17th Century.* Ann Arbor: University of Michigan Press, 1950.
Beaud, M. *A History of Capitalism.* London: Macmillan, 1984.
Beckett, J. V. *The Agricultural Revolution.* Oxford: Blackwell, 1990.
Berg, M. *The Age of Manufactures 1700–1820.* Oxford: Blackwell, 1985.
Brenner, R. "The Agrarian Roots of European Capitalism." *Past and Present* 97 (1982).
Burke, Edmund. *Inquiry into the Origin of our Ideas of the Sublime and Beautiful.* 1756.
Cipolla, C. M., ed. *The Industrial Revolution 1700–1914.* Vol. 3, *Fontana Economic History of Europe.* Brighton: Harvester Press, 1976.
Clapham, J. H. *The Bank of England.* Cambridge: Cambridge University Press, 1944.
Connolly, S. J. *Religion, Law and Power (Protestant Ireland 1660–1760).* Oxford: Clarendon Press, 1992.
Green, E. *Banking: An Illustrated History.* London: Phaidon, 1989.
Hammond, J. L., and B. Hammond. *The Village Labourer, 1760–1832.* London: Longman & Co., 1911.
Hill, C. *Society and Puritanism in Pre-Revolutionary England.* Harmondsworth: Penguin, 1986.
Jack, A. F. *Introduction to the History of Life Assurance.* London: P. S. King and Son, 1912.
Kindelberger, C. P. *A Financial History of Western Europe.* London: Allen & Unwin, 1985.
Mathias, P. *The Transformation of England.* London: Methuen, 1979.
Mingay, G. E. *A Social History of the English Countryside.* London: Routledge & Kegan Paul, 1990.
Morgan, J. *Godly Learning.* Cambridge: Cambridge University Press, 1986.
Saunders, A. Carr, and P. A. Wilson. *The Professions.* London: F. Cass, 1964.
Tawney, R. H. *Religion and the Rise of Capitalism.* Harmondsworth: Penguin, 1984.
Troeltsch, E. *Protestantism and Progress.* Pennsylvania: Fortress Press, 1986.
Weber, M. *The Protestant Ethic and the Spirit of Capitalism.* London: Allen & Unwin, 1976.
Webster, C. *The Great Instauration.* London: Duckworth, 1975.
Zaret, D. *The Heavenly Contract.* Chicago: University of Chicago Press, 1985.

Chapter 8: Class Act

Adas, M. *Machines as the Measure of Man.* Ithaca: Cornell University Press, 1985.
Adorno, T., and M. Horkheimer. *Dialectic of Enlightenment.* London: Verso Editions, 1979.

Alatas, S. H. *The Myth of the Lazy Native.* London: F. Cass, 1977.
Andrew, C. M., and A. Kanya-Forstner. *France Overseas.* London: Thames and Hudson, 1981.
Armytage, W. H. G. *Heavens Below: Utopian Experiments in England, 1560–1960.* London: Routledge & Kegan Paul, 1961.
Aziz, K. K. *The British in India.* Islamabad: National Commission on Historical and Cultural Research, 1976.
Babbage, Charles. *On the Economy of Machinery and Manufactures.* London, 1832.
Bailey, P. *Leisure and Class in Victorian England.* London: Routledge & Kegan Paul, 1978.
Baines, Edward. *History of the Cotton Manufacture of Great Britain.* 1835.
Bamford, Samuel. *Passages in the Life of a Radical.* 1893.
Bantock. A. *Studies in the History of Educational Theory.* London: Allen & Unwin, 1980.
Bastiat, F. *Cobden et la Ligue.* Paris, 1846.
Beaud, M. *A History of Capitalism.* London: Macmillan, 1984.
Beecher, J., and R. Bienvenue, eds. *The Utopian Vision of Charles Fourier.* Boston: Beacon Press, 1971.
Bellamy, Edward. *Looking Backward, 2000–1887.* Boston: Ticknor & Co., 1888.
Berard, V. *L'Angleterre et l'imperialisme.* Paris, 1900.
Berg, Maxine. *The Machinery Question and the Making of Political Economy 1815–1848.* Cambridge: Cambridge University Press, 1980.
Boahen, A. A. *African Perspectives on Colonialism.* London: Villiers, 1987.
———, ed. *Africa Under Colonial Rule 1880–1935.* Geneva: UNESCO, 1990.
Boralevi, C. *Bentham and the Oppressed.* Berlin: Walter de Gruyter, 1984.
Bradley, I. *The Call to Seriousness.* London: Jonathen Cape, 1976.
Braverman, H. "Labor and Monopoly Capital: The Degradation of Work in the Twentieth Century." *Monthly Review* (1974).
Bray, John Francis. *Labour's Wrongs and Labour's Remedy: Or the Age of Might and the Age of Right.* 1839.
Bryher, S. *The Labour and Socialist Movement Bristol.* Bristol: *Socialist Weekly* (1929).
Buchanan, R. A. *The Power of the Machine.* London: Viking, 1992.
Burke, Peter. *Popular Culture in Early Modern Europe.* London: Temple Smith, 1978.
Burnham, James. *The Managerial Revolution.* New York: J. Day & Co., 1941.
Butt, J. *Robert Owen.* Newton Abbot: David & Charles, 1971.
Cavenagh, F. A., ed. *James and John Stuart Mill on Education.* Cambridge: Cambridge University Press, 1931.
Checkland, S. G. *The Rise of Industrial Society in England, 1815–1885.* London: Longman, 1964.
Cohen, W. B. *The French Encounter with Africans.* Bloomington: Indiana University Press, 1980.
Crosland, M. P. *Science Under Control.* Cambridge: Cambridge University Press, 1992.
Curtin, P. D. *The Image of Africa.* London: Macmillan, 1965.
Da Silva, C. R., and H. W. Cane. *The Portuguese in Ceylon.* Columbo: H. W. Cane & Co., 1972.
Derry and Williams. *A Short History of Technology.* New York & London: Constable, 1993.
Digby, A., and P. Searby. *Children, School and Society in Nineteenth Century England.* London: Macmillan, 1981.
Dobbs, A. E. *Education and Social Movements 1700–1850.* London: Longman, 1919.
Eldridge, C. C. *Victorian Imperialism.* London: Hodder & Stoughton, 1978.

Foster, Lawrence. *Religion and Sexuality.* Oxford: Oxford University Press, 1981.
Gann, L. H., and P. Duignan, eds. *Colonialism in Africa.* Cambridge: Cambridge University Press, 1965.
Goodwin, B., and K. Taylor. *The Politics of Utopia.* London: Hutchinson, 1982.
Goodwin, Barbara. *Social Science and Utopia: Nineteenth-century Models of Social Harmony.* Sussex: Harvester, 1978.
Goonatilake, S. *Crippled Minds.* New Delhi: Vikas, 1982.
Hall, B. T. *Our Fifty Years.* London, 1912.
Haller, J. "The Species Problem: Nineteenth-century Concepts of Racial Inferiority in the Origin of Man Controversy." *Am. Anth.* 72(2), (1970).
Helmstadter, R. J., and P. T. Phillips. *Religion in Victorian Society.* Lantham: University Press of America, 1985.
Hempenstall, P., and N. Rutherford. *Protest and Dissent in the Colonial Pacific.* University of the South Pacific, 1987.
Hertzka, Theodor. *Freeland: A Social Anticipation.* London: Chatto & Windus, 1890.
Hollis, Patricia. *The Pauper: A Study in Working-class Radicalism of the 1830s.* Oxford: Oxford University Press, 1970.
Houghton, W. *The Victorian Frame of Mind.* New Haven: Yale University Press, 1957.
Hudson, P. *The Industrial Revolution.* London: Edward Arnold, 1992.
Hulme, P., and L. Jordanova. *The Enlightenment and Its Shadows.* London: Routledge & Kegan Paul, 1990.
Huxley, Aldous. *Brave New World.* London: Chatto & Windus, 1932.
———. *Island.* London: Chatto & Windus, 1962.
Jackson, Holbrook, ed. *On Art and Socialism.* London: John Lehmann, 1947.
Jennings, F. *The Invasion of America.* Chapel Hill: University of North Carolina Press, 1976.
Laqueur, T. W. *Religion and Respectability.* New Haven: Yale University Press, 1976.
Ludlow, J. M., and L. Jones. *Progress of the Working Class, 1832–1867.* London, 1867.
McCann, P. *Popular Education and Socialisation in the Nineteenth Century.* London: Methuen & Co., 1977.
McG. Eagear, W. *Making Men.* London: University of London Press, 1953.
Mangan, J. A. *Athleticism in the Victorian Public School.* Cambridge: Cambridge University Press, 1981.
Mangan, J. A., and J. Walvin, eds. *Manliness and Morality.* Manchester: Manchester University Press, 1987.
Mann, T., and B. Tillett. *The "New" Trade Unionism.* London: Green & McAllan, 1890.
Manuel, F. E., and F. P Manuel. *Utopian Thought in the Western World.* Oxford: Blackwell, 1979.
Moore, W. E. *Industrialization and Labor.* Ithaca: Cornell University Press, 1951.
Morton, A. L. *Life and Ideas of Robert Owen.* London: Lawrence & Wishart, 1962.
Negley, G., and J. M. Patrick. *The Quest for Utopia.* College Park, MD: McGrath Publishing Co., 1971.
"An Old Potter." *When I Was a Child.* London: Methuen, 1903.
Orwell, George. *1984.* New York: New American Library, 1962.
Peckham, H., and C. Gibson. *Attitudes of Colonial Powers Towards the American Indian.* Salt Lake City: University of Utah Press, 1969.
Pelling, H. *The Origins of the Labour Party.* New York: New American Library, 1962.
Persell, S. M. *The French Colonial Lobby.* Stanford: Hoover Institute Press, 1983.
Raskin, J., ed. *The Mythology of Imperialism.* New York: Random House, 1971.
Ridley, H. *Images of Imperial Rule.* London: Croom Helm, 1983.

Robinson, R., and J. Gallagher. *Africa and the Victorians.* London: Macmillan, 1981.
Schneider, W. H. *An Empire for the Masses.* Connecticut: Greenwood Press, 1982.
Sewell, W. H. *Work and Revolution in France.* Cambridge: Cambridge University Press, 1980.
Silver, Harold. *English Education and the Radicals 1780–1850.* London: Freedom Press, 1975.
Simon, Brian. *Education and the Labour Movement, 1870–1920.* London: Lawrence & Wishart, 1965.
Skinner, B. F. *Walden Two.* New York: Macmillan, 1948.
Smith, F. B. *Radical Artisan: William James Linton 1812–97.* Manchester: Manchester University Press, 1973.
Smith, J. M. *Seventeenth Century America.* Chapel Hill: University of North Carolina Press, 1959.
Steadman Jones, G. *Outcast London.* Oxford: Clarendon Press, 1971.
Stern, S. J. *Peru's Indian People and the Challenge of Spanish Conquest.* Madison: University of Wisconsin Press, 1982.
Suret, J., and C. Anale. *French Colonialism in Tropical Africa.* Guildford: Billings and Sons, 1976.
Tamke, S. S. *Make a Joyful Noise Unto the Lord.* Columbus: Ohio University Press, 1978.
Taylor, Barbara. *Eve and the New Jerusalem: Socialism and Feminism in the Nineteenth Century.* London: Virago, 1983.
Taylor, Frederick Winslow. *The Principles of Scientific Management.* New York & London: Harper & Bros., 1911.
Tholfsen, Trygve. *Working Class Radicalism in Mid-Victorian England.* London: Croom Helm, 1976.
Thomas, John L. *Alternative America: Henry George, Edward Bellamy, Henry Demarest Lloyd and the Adversary Tradition.* Cambridge & London: Belknap, 1982.
Thompson, Dorothy. *The Early Chartists.* London: Macmillan, 1971.
Thompson, F. M. *The Rise of Respectable Society.* London: Fontana, 1988.
Townsend, M. E. *Origins of Modern German Colonialism.* New York: Columbia University Press, 1921.
Walvin, J. *Victorian Values.* London: Cardinal, 1988.
Washburn, W. E. *Red Man's Land, White Man's Law.* New York: C. Scribner & Sons, 1971.
Wearmouth, R. F. *Some Working-Class Movements of the Nineteenth Century.* London: Epworth Press, 1948.
Wells, H. G. *When the Sleeper Wakes.* London: Harper & Bros., 1899.
———. *A Modern Utopia.* London: Chapman & Hall, 1905.
———. *Men Like Gods.* London: Cassel & Co., 1923.
———. *The Shape of Things to Come.* London: Hutchinson & Co., 1933.
Wiener, Joel. *The War of the Unstamped.* Ithaca: Cornell University Press, 1969.

Chapter 9: Doctor's Order

Ackerknecht, E. A. *A Short History of Medicine.* Baltimore: Johns Hopkins University Press, 1982.
———. *R. Virchow, Doctor, Statesman, Anthropologist.* Madison: University of Wisconsin Press, 1953.

Adorno, T. and Horkheimer. *Dialectic of Enlightenment.* Translated by J. Cumming. London: Verso Editions, 1979.

Armstrong, D. *Political Anatomy of the Body.* Cambridge: Cambridge University Press, 1983.

Bailey, P. *Leisure and Class in Victorian England.* London: Routledge & Kegan Paul, 1978.

Boyd, B. A. *R. Virchow, the Scientist as Citizen.* New York: Garland, 1991.

Brock, D. H. *The Culture of Biomedicine.* Cranberry, NJ: University of Delaware Press, 1984.

Capra, F. *The Turning Point.* London: Fontana, 1983.

Cassedy, J. H. *American Medicine and Statistical Thinking (1800–1860).* Cambridge: Harvard University Press, 1984.

Cullen, M. J. *The Statistical Movement in Early Victorian Britain.* Hassocks: Harvester, 1975.

Davis, A. B. *Medicine and Its Technology.* London: Greenwood, 1981.

Eyler, J. M. *Victorian Social Medicine.* Baltimore: Johns Hopkins University Press, 1979.

Figlio, K. "Theories of Perception and Physiology of the Mind in the Late Eighteenth Century." *Hist. Sci.* 12 (1975).

Foster, W. D. *A History of Medical Bacteriology and Immunology.* London: Heinemann, 1970.

Foucault, Michel. *The Birth of the Clinic.* London: Routledge & Kegan Paul, 1989.

———. *Discipline and Punish.* Harmondsworth: Penguin, 1979.

Greenwood, M. *The Medical Dictator.* London: Williams & Norgate, 1936.

Hacking, I. *The Taming of Chance.* Cambridge: Cambridge University Press, 1990.

Hankins, F. H. *Adolphe Quetelet as Statistician.* New York: Columbia College, 1908.

Hankins, T. *Science and the Enlightenment.* Cambridge: Cambridge University Press, 1985.

Harvey, A. *Science at the Bedside.* Baltimore: Johns Hopkins University Press, 1981.

Helmstadter, R. J., and P. T. Phillips. *Religion in Victorian Society.* Lanham: University Press of America, 1985.

Holland, W. W., R. Detel, and G. Knox. *Oxford Textbook of Public Health.* Oxford: Oxford University Press, 1984.

Hume, P., and L. Jordanova. *The Enlightenment and Its Shadows.* London: Routledge & Kegan Paul, 1990.

Jewson, N. D. *The Disappearance of the Sick Man from Medical Cosmology, 1770–1870.* London: Macmillan, 1976.

Kingsley, Charles. *Health and Education.* London: Macmillan & Co. 1882.

Kruskall, W. H., and J. M. Tanus, eds. *International Encyclopaedia of Statistics.* New York: Free Press, 1978.

McKeown, T. "Man's Health: The Past and the Future." *Western Journal of Medicine* 132 (1980).

McKinlay, J. B., and S. M. McKinlay. "The Questionable Contribution of Medical Measures to the Decline of Mortality in the United States in the Twentieth Century." *Milbank Memorial Fund Quarterly* 55 (1977).

McNeill, W. H. *Plagues and Peoples.* New York: Doubleday, 1976.

Mangan, J. A. *Athleticism in the Victorian Public School.* Cambridge: Cambridge University Press, 1981.

Mangan, J. A., and J. Walvin, eds. *Manliness and Morality.* Manchester: Manchester University Press, 1987.

Minkler, M. "People Need People: Social Support and Health." In *The Healing Brain: A*

Scientific Reader, edited by Ornstein, R., and C. Swencionis. New York: Guilford Press, 1990.
Moravia, S. "The Enlightenment and the Sciences of Man." *Hist. Sci.* 18 (1980).
Moraze, C., ed. "The Nineteenth Century." In *History of Mankind,* vol. 5, edited by C. Moraze. London: Allen & Unwin, 1976.
Muller, J. H., and B. A. Koenig. *On the Boundary of Life and Death: Biomedicine Examined.* Dordrecht: Kluwer Academic, 1988.
Ornstein, R., and D. S. Sobel. *The Healing Brain.* New York: Simon & Schuster, 1987.
Payer, L. *Medicine and Culture.* London: Victor Gollancz, 1989.
Porter, T. M. *The Rise of Statistical Thinking 1820–1900.* Princeton: Princeton University Press, 1986.
Powles, J. "On the Limitations of Modern Medicine." *Science, Medicine and Math* 1 (1973).
Reiser, J. *Medicine and the Reign of Technology.* Cambridge: Cambridge University Press, 1978.
Rhodes, P. *An Outline History of Medicine.* London: Butterworths, 1985.
Sagan, L. A. "Family Ties: The Real Reason People Are Living Longer." *The Sciences* 28 (1988).
Seligman, E., and A. Johnson, eds. *Encyclopaedia of the Social Sciences.* New York: Macmillan, 1931.
Smith, F. B. *The People's Health.* London: Weidenfeld & Nicholson, 1979.
Smith, P. *History of Modern Culture.* London: Routledge & Kegan Paul, 1930.
Starr, P. *The Social Transformation of American Medicine.* New York: Basic Books, 1982.
Sydenham, P. H. *Measuring Instruments: Tools of Knowledge and Control.* London: Peter Peregrinus, 1979.
Syme, S. L. "Sociocultural Factors and Disease Etiology." In *Handbook of Behavioral Medicine,* edited by W. Doyle Gentry. New York: Guilford Press, 1984.
Talbott, J. H. *A Biographical History of Medicine.* New York: Grune & Stratton, 1970.
Temekin, O. *The Double Face of Janus.* Baltimore: Johns Hopkins University Press, 1977.
Thompson, K. *Auguste Comte. The Foundation of Sociology.* London: Nelson, 1976.
Vigarello, G. *Concepts of Cleanliness.* Cambridge: Cambridge University Press, 1988.
Walker, K. *The Circle of Life.* London: Jonathan Cape, 1942.
Westfall, R. S. *The Construction of Modern Science.* Cambridge: Cambrige University Press, 1977.
Wohl, A. S. *Endangered Lives—Public Health in Victorian Britain.* London: J. M. Dent and Sons, 1983.
Young, G. M. *Victorian England.* Oxford: Oxford University Press, 1954.

CHAPTER 10: JOURNEY'S END

Arnon, I. *Modernisation of Agriculture in Developing Countries.* London: John Wiley, 1987.
Attewell, P. *Ground Pollution.* London: E. & F. N. Spoon, 1993.
Beddis, R. A. *The Third World.* Oxford: Oxford University Press, 1989.
Bettolo, G. Marini. *Towards a Second Green Revolution.* New York: Elsevier, 1987.
Bhagavan, M. R. *Technological Advances in the Third World.* London: Zed Books, 1990.
Bibby, C. J., et al. *ICBP: Putting Biodiversity on the Map.* Cambridge: ICBP, 1992.
Brown, Lester R., ed. *State of the World 1993.* London: Earthscan, 1993.

Chadwick, M. J., and M. Hutton, eds. *Acid Deposition in Europe.* Stockholm: Stockholm Environmental Institute, 1991.

Conway, G. R., and E. B. Barbier. *After the Green Revolution.* London: Earthscan, 1990.

Cooper, R. N. *Economic Stabilisation and Debt in Developing Countries.* Cambridge: MIT Press, 1992.

Currie, T., and A. Pepper. *Water and the Environment.* Chichester: Ellis Horwood, 1993.

Dahlberg, K. A. *Beyond the Green Revolution.* New York: Plenum, 1979.

Dogra, B. *The Greater Green Revolution.* New Delhi: Dogra Publications, 1983.

Dugan, P. *Wetland Conservation.* Switzerland: IUCN, 1990.

Elder, D., and J. Pernetta. *Oceans.* London: IUCN/Mitchell Beazely, 1991.

Foster, P. *The World Food Problem.* Colorado: Lynne Reine, 1992.

GESAMP. *The State of the Marine Environment.* Oxford: Blackwell, 1990.

Giuliano, V. "The Mechanization of Office Work." *Scientific American* 247 (1982).

Glaeser, B. *The Green Revolution Revisited.* London: Allen & Unwin, 1987.

Goldsmith, O. *5000 Days to Save the Planet.* London: Hamlyn, 1990.

Grainger, A. *Controlling Tropical Deforestation.* London: Earthscan, 1993.

Hansra, B. S. *Social, Economic and Political Implications of the Green Revolution in India.* New Delhi: Classical Publishing, 1991.

Harrison, R. M., ed. *Understanding Our Earth.* Cambridge: Royal Society of Chemistry, 1992.

Jones, G. W., ed. *The Demographic Transition in Asia.* Singapore: Maruzen Asia, 1987.

Lappe, F. M., and J. Collins. *Food First.* Boston: Houghton Mifflin, 1977.

Leon, G., and D. Hinrichsen, eds. *Atlas of the Environment 1992.* WWF, 1992.

Lindsay, P. H., and D. A. Norman. *Human Information Processing.* 2d ed. New York: Academic Press, 1977.

Marini-Betolo, G. *Towards a Second Green Revolution.* Amsterdam: Elsevier, 1987.

Mason, B. J. *Acid Rain—Causes and Effects on Inland Waters.* Oxford: Clarendon Press, 1992.

Meadows, D., and J. Randers. *Beyond the Limits.* London: Earthscan, 1992.

Meier, G. *Leading Issues in Economic Development.* Oxford: Oxford University Press, 1989.

Mercer, A. *Disease, Mortality and Population Transition.* Leicester: Leicester University Press, 1990.

Merchant, C. *The Death of Nature.* New York: Harper and Row, 1980.

Middleton, N. *Desertification.* Oxford: Oxford University Press, 1991.

Morley, L., ed. *NSCA Pollution Handbook.* Brighton: NSCA, 1992.

Myers, N. *Deforestation Rates in Tropical Forests and Their Atmospheric Implication.* London: FOE, 1991.

Neal, P. *Conservation 2000.* London: Batsford, 1993.

Newson, M., ed. *Managing the Human Impact on the Natural Environments.* New York: Belhaven Press, 1992.

Pimentel, D. *World Soil Erosion and Conservation.* Cambridge: Cambridge University Press, 1993.

Salam, Abdus. *Notes on Science, Technology and Science Education in the Development of the South.* (Privately printed.) Prepared for the South Commission, 1989.

Singh, A., and R. Singh. *The Green Revolution.* New Delhi: Harman, 1990.

Singh, J. *Dynamics of Agricultural Change.* Oxford: IBH, 1986.

Siva, V. *The Violence of the Green Revolution.* London: Zed Books, 1991.

Stone, P. B., ed. *The State of the World's Mountains.* London: Zed Books, 1992.

Teitelbaum, M. S. *The British Fertility Decline.* Princeton: Princeton University Press, 1984.
Thomson, J. *Ground Level Ozone in Canada.* Ottawa: Canadian Environment Ministry, 1992.
Tickle, A., and J. Sweet. *Critical Loads and UK Air Pollution Policy.* London: FOE, 1993.
Tolba, M., and A. El-Kholy. *The World Environment.* London: Chapman & Hall, 1992.
Tudge, C. *Food For the Future.* Oxford: Blackwell, 1988.
UNEP. *World Atlas of Desertification.* London: Edward Arnold, 1992.
WCMC Report. *Global Biodiversity—Status of the Earth's Living Resources.* London: Chapman and Hall, 1992.
Welsh, B. W., and P. Butorin, eds. *Dictionary of Development.* Vol. 1. London: St James Press, 1990.
Whitmore, T., and J. Sayer. *Tropical Deforestation and Species Extinction.* London: Chapman Hall, 1992.
WHO. *Urban Air Pollution in Megacities of the World.* Oxford: Blackwell, 1992.
Winston, P. H. *Artificial Intelligence.* Reading: Addison-Wesley, 1984.
Winston, P. H., and R. H. Brown. *Artificial Intelligence: An MIT Perspective.* Cambridge: MIT Press, 1979.
World Energy Conference. *World Energy Resources 1985–2020.* Guildford: Science and Technology Press, 1978.
Xizhe, P. *The Demographic Transition in China.* Oxford: Clarendon Press, 1991.

Chapter 11: Forward to the Past

Alexander, Michael. "X-Rays Shrink Microcircuit Lines." *Popular Science* (March 1992).
Adas, M. *Machines as the Measure of Man.* Ithaca: Cornell University Press, 1985.
Andrieu, Michel, Wolfgang Michalski, and Barrie Stevens. "Long-term Prospects for the World Economy." OECD (1992).
Appleby, Joyce. *Capitalism and a New Social Order.* New York: New York University Press, 1984.
Arnold, Guy. *The End of the Third World.* Bastingstoke: Macmillan, 1993.
Arterton, F. C. *Teledemocracy: Can Technology Protect Democracy?* London: Sage, 1987.
Attali, Jacques, and Yves Stourdze. "The Birth of the Telephone and Economic Crisis: the Slow Death of the Monologue in French Society." In *The Social Impact of the Telephone,* edited by Ithiel de S Pool. Cambridge: MIT Press, 1977.
Barnes, B., and B. J. Shapiro. *Natural Order.* Beverly Hills & London: Sage, 1979.
Bates, D. G., and F. Plog. *Social Anthropology.* New York: McGraw Hill, 1990.
Bell, Daniel. *The Coming of Post-Industrial Society: A Venture in Social Forecasting.* New York: Basic Books, 1973.
———. *Cultural Contradictions of Capitalism.* New York: Basic Books, 1979.
Berman, M. *The Reenchantment of the World.* Ithaca: Cornell University Press, 1981.
Boas, Franz. *The Mind of Primitive Man.* New York: Macmillan & Co., 1938.
Borgstrom, George. *The Hungry Planet: The Modern World at the Edge of Famine.* New York: Macmillan & Co., 1965.
Braverman, Harry. *Labour and Monopoly Capital.* London: Monthly Review Press, 1974.
Brown, Seyom. *The Causes and Prevention of War.* New York: St. Martin's Press, 1987.
Browning, John. "Libraries Without Walls for Books Without Pages: Electronic Libraries and the Information Economy." *Wired* (January 1993).

Brzezinski, Zbigniew. *The Grand Failure: The Birth and Death of Communism in the Twentieth Century.* New York: Charles Scribner's Sons, 1989.
Burke, Gerald, and Russell W. Rumberger, eds. *The Future Impact of Technology on Work and Education.* London: Falmer, 1987.
Burnham, David. *The Rise of the Computer State.* London: Weidenfeld & Nicholson, 1983.
Callenbach, Ernest. *Ecotopia.* Berkeley: Banyan Tree Books, 1975.
Calleo, David. *Beyond American Hegemony: The Future of the Western Alliance.* Brighton: Wheatsheaf, 1987.
Carey, John, and John Quirk. "The History of the Future." In *Communications Technology and Social Policy,* edited by George Gerbner. New York: John Wiley & Sons, 1973.
Carrithers, M. *Why Humans Have Cultures.* Oxford: Oxford University Press, 1992.
Chandler, Daniel. *Young Learners and the Microcomputer.* Milton Keynes: Open University Press, 1984.
Charles, Dan. "Deep Rumblings from Little Rock." *New Scientist* (December 1992).
Cheater, A. *Social Anthropology.* London: Unwin Hyman. 1989.
Chomsky, Noam. *Towards a New Cold War.* London: Sinclair Brown, 1982.
———. *The Culture of Terrorism.* London: Pluto, 1988.
———. *Necessary Illusions: Thought Control in Democratic Societies.* London: Pluto, 1989.
Clarke, Robin, and Geofrey Hindley. *The Challenge of the Primitives.* London: Jonathan Cape, 1975.
Clery, Daniel. "Memorable Future for the Lone Electron." *New Scientist* (February 1992).
———. "Will Virtual Travel Get Off the Ground?" *New Scientist* (March 1993).
Cohen, J. *The Privileged Ape.* Lancashire: Parthenon Press, 1989.
Cohen, Philip. "Against the New Vocationalism." In *Schooling for the Dole? The New Vocationalism,* edited by I. Bates. Basingstoke: Macmillan, 1984.
Connor, Steven. *Postmodernist Culture.* Oxford: Blackwell, 1989.
Countryman, Edward. *The American Revolution.* London: Taurus, 1985.
Crozier, M. P., S. J. Huntingdon, and J. Watanui. *The Crisis of Democracy: Report on the Governability of Democracies to the Trilateral Commission.* New York: New York University Press, 1975.
Dascal, M., and O. Gruengard. *Knowledge and Politics.* San Francisco: Westview Press, 1989.
Davidson, James Dale, and William Rees-Mogg. *The Great Reckoning.* London: Sidgwick & Jackson, 1993.
Davis, Frederic E. "Electrons or Photons? Read this Before You Bet on the Outer Limits of Computing." *Wired* (January 1993).
———. "Surrender the Pink!" *Wired* (May/June 1993).
Dawson, Christopher. *The Making of Europe.* London: Sheed & Ward, 1932.
De Chardin, Teilhard. *Man's Place in Nature.* London: Collin, 1956.
De Landa, Manuel. *War in the Age of Intelligent Machines.* Cambridge: MIT Press, 1991.
Demac, D. *Liberty Denied: The Current Rise of Censorship in America.* New York: Pen American Center, 1988.
DePorte, A. W. *Europe Between the Super-Powers: The Enduring Balance.* New Haven: Yale University Press, 1986.
Dickson, David. *Alternative Technology and the Politics of Technological Change.* London: Fontana, 1974.
Douglas, Mary. *Purity and Danger.* London: Routledge & Kegan Paul, 1966.
Downing, John D. H. "Computers for Political Change: PeaceNet and Public Data

Access." In *The Information Gap: How Computers and Other New Communication Technologies Affect the Social Distribution of Power,* edited by Siefert, M., G. Gerbner, and J. Fisher. Oxford: Oxford University Press, 1989.
Drexler, Eric K. *Engines of Creation.* London: Fourth Estate, 1990.
Dreyfus, Hubert L., and Stuart E. Dreyfus. *Mind Over Machine: The Power of Human Intuition and Expertise in the Era of the Computer.* Oxford: Blackwell, 1986.
———. "What Artificial Experts Can and Cannot Do." *AI and Society* (1992).
Drucker, Peter. *Postcapitalist Society.* Oxford: Butterworth Heinemann, 1993.
Edwards, Mark. "It's the Thought that Counts. Have Computers Usurped our Ability to Think?" *The Sunday Times.* London, 1 August 1993.
Ehrlich, Paul, and Anne Ehrlich. *The Population Bomb.* London: Hutchinson, 1990.
Ember, C., and M. Ember. *Cultural Anthropology.* Englewood Cliffs: Prentice Hall, 1985.
Feyerabend, P. *Against Method.* London: NLB, 1975.
Finlay, Marike. *Powermatics: A Discursive Critique of New Communications Technologies.* London: Routledge & Kegan Paul, 1987.
Firth, Raymond. *We, The Tikopia.* London: Allen & Unwin, 1936.
Fischer, Claude S. *America Calling: A Social History of the Telephone to 1940.* Berkeley: University of California Press, 1992.
Fisher, Arthur. "Crisis in Education Part III." *Popular Science* (October 1992).
Fisher, Franklin M., James W. McKie, and B. Richard. *IBM and the US Data Processing Industry.* New York: Praeger, 1983.
Florida, Richard, and Martin Kenney. "The New Age of Capitalism: Innovation-mediated Production." *Futures* (July/August 1993).
Forester, Tom, ed. *The Microelectronics Revolution.* Oxford: Blackwell, 1980.
Forsyth, Richard. *Expert Systems.* London: Chapman & Hall, 1989.
Fukuyama, Francis. *The End of History and the Last Man.* New York: Free Press, 1992.
Fuller, S. *Social Epistemology.* Bloomington: Indiana University Press, 1988.
Funtowicz, Silvio O., and Jerome R. Ravetz. "Science for the Post-normal Age." *Futures* (September 1993).
Gaddis, John Lewis. *The United States and the End of the Cold War.* Oxford: Oxford University Press, 1992.
Galbraith, John Kenneth. *The Culture of Contentment.* London: Sinclair-Stevenson, 1992.
Gale, David. "Meeting in a Virtual World." *New Scientist* (March 1993).
Gardner, Howard. *Frames of Mind.* New York: Basic Books, 1983.
Geake, Elisabeth. "Network Set to Put Journals in the Picture." *New Scientist* (November 1992).
Gehlen, Arnold. *Man in the Age of Technology.* New York: Columbia University Press, 1980.
Gennep, Arnold van. *The Rites of Passage.* London: Routledge & Kegan Paul, 1960.
Gilder. George. *Life after Television.* New York: W. W. Norton, 1992.
Goodwin, B. *Social Science and Utopia.* Atlantic Highlands: Humanities Press, 1978.
Gregory, Frank, ed. *Information Technology: The Public Issues.* Manchester: Manchester University Press, 1989.
Halal, W. "World 2000. An International Planning Dialogue to Help Shape the New Global System." *Futures* 25 (1), (1993).
Hammond, Peter B., ed. *Cultural and Social Anthropology, Selected Readings.* London: Collier Macmillan, 1969.
Harrington, Michael. *Socialism: Past and Future.* London: Pluto, 1993.
———. *The Twilight of Capitalism.* New York: Simon & Shuster, 1976.
Haviland, W. A. *Cultural Anthropology.* New York: Holt Rinehart and Winston, 1987.

Heidegger, Gerald. "Machines, Computers, Dialectics: A New Look at Human Intelligence." *AI & Society* 6, no. 1, (1992).
Herring, Ronald J. *Land to the Tiller. The Political Economy of Agrarian Reform in South Asia.* Oxford: Oxford University Press, 1983.
Hill, Christopher. *The World Turned Upside Down.* Harmondsworth: Penguin, 1984.
Hillis, Dan. *Connection Machine.* Cambridge: MIT Press, 1985.
Hoebel, E. Adamson. *The Law of Primitive Man.* Cambridge: Harvard University Press, 1956.
Holderness, Mike. "Time to Shelve the Library?" *New Scientist* (February 1992).
Ihonvbere, Julius O. "The Third World and the New World Order in the 1990s." *Futures* 24 (December 1992).
James, Clifford. *The Predicament of Culture.* Cambridge: Harvard University Press, 1988.
Jantsch, E. *Self-Organizing Universe.* New York: Pergamon, 1980.
Jeffrey, Robin. *Politics, Women and Well-being: How Kerala Became a Model.* Basingstoke: Macmillan, 1992.
Jones, G. W. *The Demographic Transition in Asia.* Singapore: Maruzen Asia, 1987.
Kahn, Herman. *The Next 200 Years: A Scenario for America and the World.* New York: Morrow, 1976.
Kahn, Herman, and Julian L. Simon. *The Resourceful Earth: A Response to Global 2000.* Oxford: Blackwell, 1984.
Kahneman, D., P. Slovic, and A. Tversky, eds. *Judgment Under Uncertainty.* New York: Cambridge University Press, 1982.
Kay, Alan C. "Computers, Networks and Education." *Scientific American* (September 1991).
Kennedy, Paul. *The Rise and Fall of the Great Powers: Economic Change and Military Conflict from 1500 to 2000.* New York: Random House, 1987.
———. *Preparing for the 21st Century.* London: HarperCollins, 1993.
Keyes, Robert W. "The Future of the Transistor." *Scientific American* (June 1993).
Kieve, Jeffrey. *The Electric Telegraph.* Newton Abbot: David & Charles, 1973.
Kleiner, Art. "If Your Toaster Had a Brain." *Wired* (January 1993).
Koestler, A. *Janus.* London: Hutchinson, 1978.
Kuhn, T. S. *The Structure of Scientific Revolutions.* Chicago: University of Chicago Press, 1962.
Langreth, Robert. "Why Scientists Are Thinking Small." *Popular Science* (April 1993).
Lasch, Christopher. *The Minimal Self: Psychic Survival in Troubled Times.* New York: Norton, 1984.
Latane, B. "The Psychology of Social Impact." *American Psychologist* 36 (1981).
Latane, B., K. Williams, and S. Hawkins. "Many Hands Make Light Work: The Causes and Consequences of Social Loafing." *Journal of Personality and Social Psychology* 37 (1979).
Lee, Dorothy. *Freedom and Culture.* Englewood Cliffs: Prentice Hall, 1959.
Levi, Werner. *The Coming End of War.* Beverly Hills & London: Sage, 1981.
Levy, Steven. "Crypto Rebels." *Wired* (May/June 1993).
Lewis, I. M. *Social Anthropology in Perspective.* Cambridge: Cambridge University Press, 1976.
Lienhardt, Godfrey. *Divinity and Experience: The Religion of the Dinka.* Oxford: Clarendon Press, 1961.
Llewellyn, K. N., and E. Adamson Hoebel. *The Cheyenne Way: Conflict and Case Law in Primitive Jurisprudence.* Oklahoma: University of Oklahoma Press, 1941.

Long, Chris, and Claire Neesham. "Whose Copyright Is It Anyway?" *New Scientist* (July 1993).

Longuet-Higgins, Christopher. *Mental Processes.* Cambridge: MIT Press, 1987.

Lowie, Robert H. *Indians of the Plains.* New York: McGraw Hill Book Co., 1954.

———. *Social Organizations.* London: Routledge & Kegan Paul, 1950.

Lyotard, Jean-François. *La Condition Postmoderne: Rapport sur le Savoir.* Paris: Les Editions de Minuit, 1979.

Malinowski, Bronislaw. *Crime and Custom in Savage Society.* London: Kegan Paul & Co., 1926.

Malone, Thomas W., and John F. Rockart. "Computers, Networks and the Corporation." *Scientific American* (September 1991).

Manganaro, M. *Modern Anthropology.* Princeton: Princeton University Press, 1990.

Mankekar, D. R. *The Red Riddle of Kerala.* Bombay: Manaktalas, 1965.

Marcuse, Herbert. *Eros and Civilization.* London: Routledge & Kegan Paul, 1955.

———. *One Dimensional Man.* London: Routledge & Kegan Paul, 1964.

Marglin, S. A. "Knowledge and Power." In *Firms, Organization & Labour,* edited by F. H. Stephen. London: Macmillan, 1984.

Marvin, Carolyn. *When Old Technologies Were New: Thinking about Electric Communication in the Nineteenth Century.* Oxford: Oxford University Press, 1988.

Maxwell, N. *From Knowledge to Wisdom.* Oxford: Blackwell, 1984.

Mbiti, John S. *African Religions and Philosophy.* London: Heinemann, 1969.

Mead, Margaret. *Sex and Temperament in Three Savage Tribes.* London: G. Routledge & Sons, 1935.

Mearsheimer, John. "Why We Will Soon Miss the Cold War." *The Atlantic* (August 1990).

Meier, G. *Leading Issues in Economic Development.* Oxford: Oxford University Press, 1989.

Mercer, A. *Disease, Mortality and Population in Transition.* Leicester: Leicester University Press, 1990.

Michie, Donald, and Rory Johnston. *The Creative Computer: Machine Intelligence and Human Knowledge.* London: Viking, 1984.

Miles, Ian, et al. *Information Horizons: The Long-term Social Implications of New Information Technologies.* Aldershot: Elgar, 1988.

Moravec, Hans. *Mind Children.* Cambridge: Harvard University Press, 1988.

Morris, M. D. *Measuring the Condition of the World's Poor: the Physical Quality of Life Index.* New York: Pergamon, 1979.

Morris-Suzuki, Tessa. *Beyond Computopia: Information, Automation and Democracy in Japan.* London: Kegan Paul, 1988.

Moser, P., and Nat A. Vanden. *Human Knowledge.* Oxford: Oxford University Press, 1987.

Mowlana, Hamid. *Global Information and World Communication: New Frontiers in International Relations.* New York: Longman, 1986.

Mumford, Lewis. *Technics and Civilization.* New York: Harcourt Brace, 1963.

Myers, Norman, ed. *Gaia: An Atlas of Planet Management.* Garden City: Anchor Press, 1984.

Naisbitt, John, and Patricia Aburdene. *Megatrends 2000.* London: Sidgwick & Jackson, 1990.

Nester, William R. *American Power, the New World Order and the Japanese Challenge.* Basingstoke: Macmillan, 1993.

Noble, David. *The Forces of Production.* Oxford: Oxford University Press, 1986.

O'Brien, Richard. *Global Financial Integration: The End of Geography.* London: Pinter, 1992.

O'Connor, D. J., and B. Carr. *Introduction to the Theory of Knowledge.* Brighton: Harvester Press, 1982.

OEDC Forum for the Future. *Long-term Prospects for the World Economy.* Paris, 1992.

Ohmae, Kenichi. *The Borderless World.* New York: Collins, 1990.

Pacey, Arnold. *The Cult of Technology.* Cambridge: MIT Press, 1983.

———. *The Maze of Ingenuity.* Cambridge: MIT Press, 1992.

Paddock, William, and Paul Paddock. *Famine 1975!* Boston: Little, Brown, 1967.

Papert, Seymour. "Computer and Learning." In *The Computer Age: A Twenty-Year View,* edited by Dertouzos, M. L., and Joel Moses. Cambridge: MIT Press, 1979.

———. *Mindstorms: Children, Computers and Powerful Ideas.* New York: Basic Books, 1980.

———. "Literacy and Letteracy in the Media Ages." *Wired* (May/June 1993).

Paterson, Thomas. *Meeting the Communist Threat.* Oxford: Oxford University Press, 1988.

Paul, Benjamin D. "Mental Disorders and Self-Regulating Processes in Culture: A Guatemalan Illustration." In *Personalities and Cultures: Readings in Psychological Culture,* edited by Robert Hunt. New York: Natural History Press, 1967.

Perelman, Lewis. "School's Out: Public Education Obstructs the Future." *Wired* (January 1993).

Phillips, D. P. "The Impact of Mass Media Violence on U.S. Homicides." *American Sociological Review* 48 (1983).

Pool, Ithiel de S. *The Social Impact of the Telephone.* Cambridge: MIT Press, 1977.

———. *Forecasting the Telephone.* Norwood: Ablex Publishing, 1983.

———. *Technologies of Freedom.* Cambridge: Belknap, 1983.

Radcliffe-Brown, A. R. *Structure and Function in Primitive Society.* London: Cohen & West, 1952.

Radin, Paul. *The World of Primitive Man.* New York: Appleton & Co., 1953.

Rappaport, Roy. *Pigs for the Ancestors.* New Haven: Yale University Press, 1968.

———. "Nature, Culture and Ecological Anthropology." In *Man, Culture and Society,* edited by H. Shapiro. Oxford: Oxford University Press, 1960.

Rasmussen, Knud. *Across Arctic America.* New York: G. P. Putnam's Sons, 1927.

Rivers, W. H. *Instinct and the Unconscious.* Cambridge: Cambridge University Press, 1924.

Robins, Kevin, and Frank Webster. *The Technical Fix: Education, Computers and Industry.* Basingstoke: Macmillan, 1989.

Roszak, Theodore. *The Cult of Information: The Folklore of Computers and the True Art of Thinking.* New York: Pantheon, 1986.

Schumacher, E. F. *Small Is Beautiful: Economics As If People Mattered.* London: Blond & Briggs, 1973.

Shapiro, H., ed. *Man, Culture and Society.* Oxford: Oxford University Press, 1971.

Simon, Julian L. *The Ultimate Resource.* Oxford: Martin Robertson, 1981.

Simons, Geoff. *Eco-Computer: The Impact of Global Intelligence.* Chichester: Wiley, 1987.

Singer, A., and L. Woodhead. *Disappearing World.* London: Boxtree, 1988.

Skeels, H. M. "Adult Status of Children with Contrasting Early Life Experience." *Monographs of the Society for Research in Child Development* 31(3), (1966).

Skinner, B. F. "What Is Wrong with Behavior in the Western World?" *American Psychologist* 41(5), (1986).

Slaughter, Richard. "Looking for the Real 'Megatrends.'" *Futures* (October 1993).

Smith, Anthony. *Goodbye, Gutenberg: The Newspaper Revolutions of the 1980s.* Oxford: Oxford University Press, 1980.

Smoke, Richard., and Willis W. Harman. *Paths to Peace: Exploring the Feasibility of Sustainable Peace.* Boulder: Westview, 1987.

Sousa, E. *Knowledge in Perspective.* Cambridge: Cambridge University Press, 1991.

South Commission. *The Challenge to the South: The Report of the South Commission.* London, 1990.

Spengler. O. *The Decline of the West.* London: Allen & Unwin, 1926.

Sproull, Lee, and Sara Kiesler. "Computers, Networks and Work." *Scientific American* (September 1991).

Stocking, G. W. *Race, Culture and Evolution.* Chicago: University of Chicago Press, 1982.

Stonier, Tom. *Beyond Information: The Natural History of Intelligence.* London: Springer-Verlag, 1992.

Stonier, Tom, and Cathy Conlin. *The Three Cs: Children, Computers and Communication.* Chichester: Wiley, 1985.

———. *The Wealth of Information: A Profile of the Post-Industrial Economy.* London: Thames Methuen, 1983.

Strathern, A. *Landmarks—Reflections on Anthropology.* Ohio: Kent State University Press, 1993.

Teitelbaum, M. S. *The British Fertility Decline.* Princeton: Princeton University Press, 1984.

Tesler, Lawrence G. "Networked Computing in the 1990s." *Scientific American* (September 1991).

Tjosvold, D. *Working Together to Get Things Done: Managing for Organization Productivity.* Lexington: Heath, Lexington Books, 1986.

Toffler, Alvin. *Future Shock.* New York: Bantam Books, 1971.

———. *The Third Wave.* New York: Bantam Books, 1981.

———. *Powershift: Knowledge, Wealth and Violence at the Edge of the 21st Century.* New York: Bantam Books, 1991.

Tversky, A. "Features of Similarity." *Psychological Review* 84 (1977).

Tversky, A., and D. Kahneman. "The Framing of Decisions and the Psychology of Choice." *Science* 211 (1981).

Tylor, Edward B. *Anthropology: An Introduction to the Study of Man and Civilization.* London: Watts & Co., 1930.

Varlaam, Carol., ed. *Rethinking Transition: Educational Innovation and the Transition to Adult Life.* London: Falmer, 1984.

Wagar, W. Warren. *Building the City of Man: Outlines of a World Civilization.* San Francisco: W. H. Freeman & Co., 1971.

———. *A Short History of the Future.* Chicago: University of Chicago Press, 1989.

———. *The Next Three Futures: Paradigms of Things to Come.* London: Adamantine Press, 1992.

Walkerdine, Valerie. "Developmental Psychology and the Child-Centred Pedagogy: The Insertion of Piaget into Early Education." In *Changing the Subject: Psychology, Social Regulation and Subjectivity,* edited by Julian Henriques. London: Methuen, 1984.

Wallerstein, Immanuel. *The Capitalist World-Economy.* Cambridge: Cambridge University Press, 1979.

———. *The Politics of the World-Economy: The States, the Movements and the Civilizations.* Cambridge: Cambridge University Press, 1984.

Weiser, Mark. "The Computer for the 21st Century." *Scientific American* (September 1991).

Weizenbaum, Joseph. *Computer Power and Human Reason.* Harmondsworth: Penguin, 1984.

Welsh, B., and P. Butorin. *Dictionary of Development.* London: St James' Press, 1990.

Winston, P. H., and K. A. Prendergast, eds. *The AI Business: Commercial Uses of Artificial Intelligence.* Cambridge: MIT Press, 1984.

Winthrop, R. H. *Dictionary of Concepts in Cultural Anthropology.* New York: Greenwood Press, 1991.

Yarrow, L. J., et al. "Dimensions of Early Stimulation and Their Differential Effects on Infant Development." *Merrill-Palmer Quarterly* 18 (1972).

Yates, S., and E. Aronson. "A Social-Psychological Perspective on Energy Conservation in Residential Buildings." *American Psychologist* 38 (1983).

Yazdani, M., ed. *Artificial Intelligence.* Chichester: Ellis Horwood, 1986.

Zuboff, S. *In the Age of the Smart Machine: The Future of Work and Power.* New York: Basic Books, 1988.

Index

Abelard, Peter, 112
Accademia del Cimento, 157
Accounting, 43–45
Act of Tolerance, 161–62
Adaptation, 6–7, 23, 289
Adelard of Bath, 111–12
Aeneid (Vergil), 94
Aeschylus, 84–85
Africa, 221–22, 267–68, 272
Agriculture, 45–46
  English improvements in, 177–84
  Green Revolution, 258–62
  primitive, 36–40
  short-fallow farming, 45
  topsoil loss, 266–67
  See also Irrigation
Air, 166
Air pollution, 264–65, 290
Air pressure, 164
Almanacs, 137–38
Al Mansur (caliph), 104
Alphabet, 58–60
  Greek, 64, 69–70
Alphabetic thinking, 64–72
  and Plato, 86–87
America, 147, 149, 289
  See also United States
Amorites (people), 55
Amsterdam (Netherlands), 188
Analysis, 81
The Anatomy of Leaves, Flowers and Fruits (Grew), 167
Anaximander, 74–75
Anaximenes, 75
Andral, Gabriel, 233
Anglican Church, 158
Aniline dyes, 197
Animal husbandry, 179–80
Anthrax, 245
Antigone (Sophocles), 85
Apartheid, 218–19
Aquinas, St. Thomas, 116–17
Arational thinking, 293
Aristotle, 74, 80–82
  ban on teaching, 115–16
  on democratic government, 301
  on earthly movement, 150
  on gravity, 151
  Great Chain of Being, 107
  philosophy of, 111
  and specialization, 84
  on vacuums, 162–63
Arithmetical signs, 44–45
Arrows, 37
Art
  beginnings of, 27–29
  as propaganda, 98–99
Ascultation, 233
Asia, 268
Astrology, 111
Astronomy, 111, 168
Aswan dam, 263
Asymmetric limbs, 9–10
Athens (Greece), 85–86
Athletics, 239–40
Atmospheric circulation, 5, 6
Atomism, 74
Auenbrugger, Leopold, 233
Augustine, St., 96–97
Austen, Ralph, 184
Australia, 256
Australopithecus, 9
Axes, 10–12

Babel, Heinrich, 133
Babylon, 55

Bacon, Francis, 153–54
Bacon, Roger, 117–18
Bacteriology, 197, 245–47, 271
Bakewell, Robert, 179–80
Ballistics, 152
Baltic Sea, 266
Bank of England, 189
Banks and banking, 188–89, 276
Barometer, 164–65
Batons, 29–33
  purpose of, 30–31
Bede, St., 99
Bell, Andrew, 205–6
Belloc, Hillaire, 287–88
Benedictines, 107
Berlin Conference, 221–22
Bernard, Claude, 246
Bernard of Clairvaux, St., 107–8
Berti, Giovanni, 163
Bible
  appendices, 135–37, 286
  King James, 127
  vernacular, 124–25, 131, 133
Bichat, Xavier, 231–32
Biography, 69
Biological diversity, 268
Biology, 168
Biometer, 238
Bipedalism, 9
Birth rates, 272, 273–74
Black Dwarf (publication), 204
Bleaching, 263
Blood, 233
Boniface, St., 98
Book of Common Prayer, 127–28
Borlaug, Norman, 260
Botanical garden, 150
Botany, 139, 167–68, 288
Boustrophedon movements, 71
Boyle, Robert, 160–61, 164, 165
Brain, 29–30
  asymmetry of, 10, 71–72
  evolution of, 13
  expansion of, 12
  and experience, 291–92
  flexibility of, 291–92
  neural connections, 16
  sequentially thinking, 18–20
  talent potential of, 309–10
Brazil, 268
British Association for the Promotion of Co-operative Knowledge, 204
British Report on Technical Instruction, 214
Broca's area, 29
Brown, Capability, 176, 182, 183
Bureaucracy
  Egyptian, 57–58
  and printing, 129
  Roman, 94
Burial practices, 26–27, 37

Cabanis, Pierre, 230
Calcutta (India), 271
Calendars, 31–32
Camden, William, 132
Canaanites, 59
Canals, 190
Capitalism, 123, 134–35, 186–93
Carbon dioxide, 269–70
Carnaervon, Earl of, 221
Caro, Heinrich, 197
Carrier bags, 17
Carrying capacity, 23, 36, 38, 258
Catholic Church
  ban on teaching of Aristotle, 115–16
  censorship by, 131
  communications network, 97–98
  consolidation of authority, 100–101, 102–3
  diminishing power of, 133–34
  and education, 112, 141
  and heliocentrism, 151
  hierarchy of, 97
  and man's dominion over nature, 106–7
  papal leadership of, 101–2
  and printing, 124
  scholarship by, 112–14
  social control by, 103–4
  and translations from Arabic, 109–11
Caxton, William, 126
Cellular pathology, 243–44
Censorship, 131–32
Chadwick, Edwin, 236–37
Chesapeake Bay, 263
Chieftainship, 255
Chigi Vase, 84
Child labor, 216
China, 105–6
Cholera, 234–35, 238–39, 240, 246, 264
Chomsky, Noam, 297
Circle of Auteuil, 230
Cistercians, 107–9
Class, 202–3, 210–11

Classification, 171–72
Clinical Medicine (Pinel), 231
Clocks, 109, 169, 195
Coal, 263, 264
Cobbett, William, 180, 181, 203
Code of Hammurabi, 55–56
Coke, Thomas, 180
Collins, John, 170
Colonialism, 218–23
 See also Electronic colonialism
Columbus, Christopher, 147
Combustion, 165
Comenius, Jan Amos, 141
Communications technology, 284, 286
Communism, 216
Compact discs, 296
Computers, 282–90, 293–96, 308
 See also Web data-links
Condorcet, Marquis de, 229–30
Confession, 103–4
Conformity. *See* Social control
Confucius, 105
Cooperative societies, 204–5
Copernicus, Nicolaus, 151
Coral reefs, 263
Cordus, Valerius, 139
Cosmology, 150–51, 289
Counting, 41–45
 one-to-one correspondence, 41–42
Crafts, 46
Cranmer, Thomas, 127–28
Cree Indians, 15
Crop rotation, 178
Cruciger, Caspar, 139
Cultural diversity, 306, 311
Cuneiform, 50

Dark Ages, 93
Data collection, 229–30
Data-processing techniques, 283–84
De Civitate Dei (St. Augustine), 96
Deductive reasoning, 81
Deforestation, 267–69
DeKerckhove, Derrick, 71
Democracy, 69, 85–86, 301–2
Democritus, 74
Demotic writing, 57
Descartes, Rene, 155–56
Detrosier, Rowland, 212–13
Dewline defense system, 283
Dictatus Papae (1075), 101
Diet, 38
Discourse on Method (Descartes), 155–56
Discourse on Trade (North), 171, 187
Disease, 226–27, 271
Division of labor, 11–12, 52, 192, 245
Donation of Constantine, 101–102
Donne, John, 153
Dry-farming, 38
Dutch East India Company, 188
Dyes, 197

Easter Island, 256
Ecole de Santé, 231
Ecology, 24–25
 See also Pollution
Economics, 187–90
 mechanistic, 171, 188
 Smith's philosophy, 172, 191–93
Edubba (scribal school), 51
Education, 140–42
 Catholic, 112, 141
 and computers, 296, 308
 Confucian, 105–6
 German, 140–41
 Greek, 70–71, 72, 76
 Japanese, 69–70
 in Kerala, India, 304–5
 in Middle Ages, 98
 scribal schools, 51
 and social control, 205–7, 211–12, 216
 technical, 213–14
 Western-style, 300
Egypt
 bureaucracy in, 57–58
 centralization of authority in, 57
 development of, 57
 scribes in, 50–51
 writing in, 56, 58–59
 See also Aswan dam; Nile River
Ehrlich, Paul, 246
Ehrlich, Paul R., 273
Electricity, 165, 263
Electronic agents, 287
Electronic colonialism, 287–88
Electronic fund transfer, 284–85
Elites, 46–49
Empedocles, 75
Enclosure, 178–79
Encyclopaedia of the Social Sciences, 297
Energy cycles, 5–7
Energy use, 264
Engineers, 214
England. *See* Great Britain

Epidemics, 243, 247–48
Erosion, 266–67
Etymologies (Isidore of Seville), 99
Euripides, 84, 85
Evangelical Movement Moral Crusade, 203, 209
Evolution, 18–19
Exchange Bank, 188
Experience. *See* Observation
Experiment, 118–19
Experimental Philosophy Club, 157–58
Extinction, 256, 268–69
Eyes
  of early hominids, 10
  nearsightedness, 15–16

"A Factory As It Might Be" (Morris), 217
Farming. *See* Agriculture
Farr, William, 238
"Father in Heav'n" (Kipling), 210
Fertility rate, 274
Fertilizer, 258, 260
Fire, 12, 263
Fishing, 262–63
Fontana, Niccolò, 151–52
Food and Agricultural Organization, 262
Food surplus, 39–40, 258
France, 129
Francis I (king of France), 131
Free association (social structure), 148–49
French West Indies, 218–19
Freshwater cycle, 6
Fuel wood, 263, 267, 268

Galbraith, John Kenneth, 305
Galilei, Galileo, 152, 159, 163, 165
Gallup, Gordon, 21
Gas, natural, 264
Gascoigne, William, 168
Gelasius I (pope), 101
General Anatomy (Bichat), 232
General Registry Office (Great Britain), 236
Georgias of Leontini, 77–79, 80
Gerard of Cremona, 110
Germania (Tacitus), 133
Germany
  education in, 140–41
  national history, 133
  national language, 126
  and statistics, 241
Girault, Arthur, 220
Global warming, 269–70
Gnosticism, 100
Grammar, 22, 126
Gravity, 152
Great Britain
  agriculture, 177–84
  canals, 190
  cholera epidemic in, 234–35, 238–40
  and India, 219–20
  national history, 132
  national language, 127–28
  population, 235
  social conditions, 237
Great Chain of Being, 82, 107
Great Plains (U.S.), 262
Greece, 66–87
  alphabet, 64, 69–70
  early literature of, 68
  political life of, 85–86
  style of thought, 75
  theater of, 84–85
Greenhouse Effect, 269–70
Green Revolution, 258–61
Gregory I (pope), 98–99
Gregory VII (pope), 101
Gregory IX (pope), 116
Grew, Nehemiah, 167–68
Grinding stones, 37
Grossteste, Robert, 118
Growth. *See* Limitless growth concept
Guicciardini, Francesco, 132–33
Gutenberg, Johann, 122–23, 311

Haeckel, Ernst, 244–45
Hall of Science (Manchester, England), 207–8
Hammurabi (king of Babylon), 55
Hand axes, 11–12
Handbook for U.S. Public Health, 248
Hands, 10
Harrison, John, 169
Harvey, William, 167
Health, 264
  See also Medicine; Public health
Heliocentrism, 151, 289
Henry IV (king of France), 125
Henry VIII (king of England), 142
Heraclitus, 73–74
Herodotus, 69
Hesiod, 84
Hieratic writing, 56–57
Hieroglyphs, 59
Historiae de rebus Hispaniae (Mariana), 133

History, 132–33
  and alphabet, 69
  Condorcet's views, 229
History of Florence (Guicciardini), 132
History of Italy (Guicciardini), 132
Hollerith, Herman, 281–82
Holy Roman Empire, 129–30
Homo erectus, 11
Homo habilis, 10–11
Homo sapiens, 20
Horace (Roman poet), 94
Hugo, Victor, 220
Humans
  adaptive abilities of, 23–24
  earliest ancestor, 8–9
  evolution of, 19
  impact on ecology, 24–25
  main physical groups of, 23
  mental processes, 280
  migration of, 22–23, 33
  See also Brain
Hunter-gatherers, 23
Hunting, 10–11, 37
  ice-age, 25
Huntsman, Benjamin, 195
Hygiene, 239
Hymns, 208–10
Hymns Ancient and Modern, 208

IBM (International Business Machines), 282
Ice Age, 25–26
Imaginative thinking, 292–93
Imitation of Christ (devotional text), 124
India, 219–20, 259, 260, 267
Industrialization, 300
  and fertility rate, 274
Industrial Revolution, 165, 194–97, 201–2, 258, 264, 269
Industrial waste dumping, 265–66
Information, 274–75
  control of public, 296–98
  increasing access to, 298
  See also Knowledge
Inheritance (genetic), 19
Institute of Civil Engineers, 214
Instruments, scientific, 168
  medical, 233–34
  standardized, 161
  See also specific instruments
Insurance companies, 189–90, 276
Intelsat 284
Interchangeability, 196
Intermediate technology, 259
An Introduction of the Eyght Partes of Speche (Lily), 128
Irrigation, 45–46, 258, 261–62, 267
Isidore of Seville, 99
Islam, 104–5, 106
Italy, 132

Jacquard loom, 282
Japanese education, 69–70
Jeffrey, Robin, 305
Jesuit colleges, 141
Johnson, Lyndon, 259
Jonathan's Coffeehouse (London), 190
Jonnes, Moreau de, 241

Kebarans (people), 37
Kent, William, 182–83
Kepler, Johann, 137
Kerala (India), 303–6
King James Bible, 127
Kingsley, Mary, 220
Kipling, Rudyard, 210
Knowledge, 60, 149, 275
  as artifact, 72–73
  and computers, 287–90, 295
  general crisis in, 153
  modern view of, 155
  relativism in, 76–78
  and secrecy, 289
  traditional indigenous, 300
  See also Information; Philosophy
Knowledge-based systems, 287
Koch, Robert, 197, 245–46
Kojalavicus, Albertas Vijukas, 133
Komensky, Amos. *See* Comenius, Jan Amos
Korea, 123

Labor
  child, 216
  conditions, 201–2
  division of, 11–12, 52, 192, 245
  in factories, 215
  Morris on, 216–17
  organization, 204
  right of movement of, 181
  as virtue, 184–85
  and wages, 191
  See also Unemployment
Labor camps, 211
Laënnec, Théophile, 233

"La Marche" baton, 31
Lancaster reading method, 206
Landscaping, 182–83
Language, 306
  consolidation of, 38
  development of, 12–13
  development of national, 125, 127
  differentiation of, 33
  modern families of, 38
  standardization of, 126
  and toolmaking, 21–22
  and truth, 77
  and vernacular Bibles, 124–25
  See also Alphabet; Grammar; Writing
Laplace, Pierre Simon, 228–29
Latitudinarians, 158
Law, 53–56
  and printing, 130
Lead pollution, 265, 290
Leakey, Mary, 9
Leeuwenhoek, Anton van, 167
Leibniz, Gottfried, 232, 307
Levallois technique, 20–21, 26
Lily, William, 128
Limitless growth concept, 299–300
Linnaeus, Carolus, 171–72
Lister, J. J., 243
Literacy
  and democracy, 86
  and Greek alphabet, 69, 70
  in Kerala, India, 305
  in Mesopotamia, 53
  in Middle Ages, 98
  in Protestant community, 132
  scribal schools, 51
  and social control, 75–76, 140
Literature, 68–69
Locke, John, 172–73, 186
Logic, 82, 87, 111
London Co-operative Society, 204
Longitude, 169
Loom, 283
Luther, Martin, 125–26, 129, 139, 140

McQueen, James, 222
Magnus, Albertus, 113–14
Magnus, Olaus, 133
Malpighi, Marcello, 167
Malthus, Thomas Robert, 273
Management, 215
Man and the Development of his Faculties: A Social Physics (Quetelet), 241
Manuals, 138
Maoris (people), 256
Maps, 275–76
Mariana, Juan de, 133
Marine chronometer, 169, 195
Marshall, Alfred, 211
Mass production, 12, 264
Mathematics, 79–80, 83
  See also Counting; Numbers; Probability mathematics
Maudslay, Henry, 196
Maximilian (Hapsburg emperor), 130–31
Maybach, William, 197
Measurement, 54, 152, 168–72
Mechanics' Institutes (Great Britain), 211–12
Medea (Euripides), 85
Media, 297
Medicine, 111, 142, 226–49
  diagnosis, 246–47
  instruments for, 233–34
  and population growth, 271, 274
  and statistics, 231, 232, 235–36
  See also Disease; Public health
Mediterranean Sea, 265
Memory, 69, 70
Mersenne, Marin, 153
Mesopotamia, 43–50, 52–55
Metaphysics (Aristotle), 74
Mexico, 259
Mexico City, 265, 271, 290
Micrometer, 168–69
Microscope, 166–68, 233, 243
Miletus (Greece), 73
Miracles, 114–15
Missionaries, 218, 220
Molière, Jean-Baptiste, 227
Monads, 232
Monasteries, 95–96
  Cistercian, 108–9
  liturgical needs of, 109
Money supply, 171
Monocultures, 261
Montesquieu, Charles Louis de Secondat, Baron de la Brède et de, 156
Montano, Benito Arias, 135, 136
Montgaudier baton, 30–31
More, Hannah, 202–3, 206
Morris, William, 216–18
Morrison, James, 208
Mortality rates, 273–74
Muller-Lyer illusion, 15

Multiple witnessing, 160
Muscular Christianity, 240
Museion (Alexandria), 83

Names, personal, 37
National identity, 125–26, 132–33, 306
Nationalism, 127, 128
Nation states, 306
Natufian culture, 37–38
The Natural History of Plants (Ray), 168
Nature, 106–7, 182, 258
Nature-philosophy movement, 231–32
Nearsightedness, 15–16
Nestorians, 104
Nestor's Cup, 69
Netherlands, 130
New Lanark (Scotland), 207
The New Public Health, 247–48
The New System (Bacon), 153
New Zealand, 256
Nile River, 57, 263
Nippur (Mesopotamia), 54–55
North, Dudley, 171, 187–88
North Sea, 265
Numbers, 41–42

Objectivity, 155, 159, 161
Observation, 152, 154
Ogallala aquifer, 262
Oil. *See* Petroleum
"On the Degree of Certainty in Medicine" (Cabanis), 230
On the Non-existence of Existence, Or on Nature (Georgias), 77
On temperature in diseases (Wunderlich), 234
Optical illusions, 15
Opus Maius (Bacon), 117
Oral culture, 67
Oral memory, 69
Oresme, Nicholas, 119
Ostracism, 86
Ovando, Juan de, 150
Owen, Robert, 207–8
Oxygen cycle, 6

Paine, Thomas, 202
Pakistan, 259–60
Pantheism, 115
Papin, Denis, 165
Papyrus, 56
Parallel processing, 285
Parmenides, 73
Pascal, Blaise, 163–64
Pathology, 232, 243–44
Pebble-axes, 10
Penny Magazine, 212
Pepys, Samuel, 170
Perception, 13–15
Percussion (medical technique), 233
Perkin, William, 197
Perry, William, 187
Persian Letters (Montesquieu), 156–57
Peter of Maricourt, 117
Petroleum, 263
  reserves, 264
Petty, William, 170
Philosophica Botanica (Linnaeus), 171–72
Philosophical Transactions of the Royal Society (journal), 160
Philosophy, 73–80
  and theology, 114, 116–17
Phoenicians, 65–66
Picard, Jean, 165, 169
Pictograms/pictographs, 51, 52, 68
Pinel, Philippe, 231
Pippin the Short, 100
Plantin, Christopher, 134–35
Plato, 77, 79–80
  and alphabetic thinking, 86–87
Ploughs, 38, 45, 258
Poaching, 180–81
Poetry, 67–68
Politics, 172
  See also Democracy
Pollution
  air, 264–65, 290
  water, 262–63, 266
Poor Man's Guardian, 212
Pope, Alexander, 176, 182
Population, 258, 270–74, 277, 303–6
Pottery, 84
Poverty, 180, 272
Power, 17, 257
  Egyptian, 58
  and literacy, 75
  and printing, 130
  tokens, 43
  See also Social control
Practical Observations on the Education of the People, 211
Printing, 123–43
  and bureaucracy, 129
  and Catholic Church, 124

of maps, 275
and power, 130
and propaganda, 130–31
Privacy, 285
Probability mathematics, 228–29, 230, 241
Pro et contra (logic method), 112
Professions, 142, 214
Propaganda
hymns for, 208–10
and printing, 130–31
Roman, 94
Property ownership, 53–54, 186
Protagoras, 76
Protestants, 132, 185
Protohorticulture, 36
Ptolemy, 150–51
Public health, 238–40, 243
laboratories, 247
and mortality rate, 273–74
social control through, 248
World War II techniques, 271
Public opinion, 299
Pulley blocks, 196
Punched-blade technique, 26
Puritans, 132, 184–85
Pygmies (people), 16

Qafzeh cave (Israel), 20
Quackelbeen, Willem, 139
Quantification, 168–72
Quetelet, Adolphe, 241–42
Quinine, 197

Racism, 220–23
Raikes, Robert, 206–7
Rain, 6
Ramdsen, Jesse, 195
Ramidus, 8–9
Random generation, 18–19
Rationalism, 116–17
Ray, John, 168, 171
Reading, 16, 126, 206, 308
Reductionism, 156, 157, 228, 288
Relativism, 76–78, 155–67
Religion, 255, 306
and nationalism, 128
and science, 158
theology, 114, 116–17
See also Catholic Church; Protestants
Remaines Concerning Britaine (Camden), 132
Residuum, 210–11
Resolution and composition, 118
Respiration, 165
Rhetoric, 77
Rhodes, Cecil, 221
Rice, 258–59, 260
Rights of Man (Paine), 202
Roman Catholic Church. *See* Catholic Church
Roman Empire, 93–95
Rousseau, Jean Jacques, 301
"Royal" Bible, 135–36
Royal College of Medicine, 142
Royal Society, 158–62, 167
Royal Society for the Encouragement of Arts, Manufacture, and Commerce, 184
Rural social structure, 215
Russia, 265

SABER (computer network), 283
Sails, 45
St. Augustine. *See* Augustine, St.
St. Bede. *See* Bede, St.
St. Boniface. *See* Boniface, St.
St. John, Henry, 190
St. Thomas Aquinas. *See* Aquinas, St. Thomas
Salford Social Institution, 208
Salinization, 267
Salvarsan, 246
Sanitation, 237
Satellites, 284
Schleiden, Matthias, 243
Schwann, Theodor, 243
Science, 70, 117, 148
and political power, 166
and religion, 158
and social control, 159
See also specific disciplines, e.g., Botany
Scientific instruments. *See* Instruments, scientific
Scientific method, 156, 157–62, 172–73
Scientific Revolution, 176, 286
Score satellite, 284
Scribal schools, 51
Scribes, 50–51
Seals (personal stamps), 50
Sea-temperature cycle, 5
Sedimentation, 263
Selective retention, 18–19
Self-help movement, 212–13
Self-Help (Smiles), 213

Self-interest, 173, 186, 193
Semliki valley (Zaire), 20
Senses, 14
Serabit el Khadem (Sinai peninsula), 58–59, 60
Settlement, 36–37, 40
Sewage systems, 237
Sextant, 169
Shaftesbury, Lord, 182
Shamans, 48
Shipley, William, 183–84
Sic et Non (Abelard), 112
Sicklemore, Samuel, 176
Sickles, 37
Smiles, Samuel, 213
Smith, Adam, 172, 191–93
Smollett, Tobias, 182
Snow, John, 240
Social class, 202–3
Social control, 28, 37, 41, 201, 299
  and Catholic Church, 103–4
  and church ceremony, 208
  and class, 202–3
  and education, 205–7, 211–12, 215
  and literacy, 75–76
  and logic, 111
  in Mesopotamia, 47–49, 60–61
  and printing, 130–31
  and public health, 248
  and science, 159
  and statistics, 229, 241–42
  in Sumeria, 310
  See also Law
Social Darwinism, 211
Social Democratic Foundation, 215
Socialism, 204–5, 207–8, 216–17
Society for Civil Engineers, 214
Society for the Diffusion of Useful Knowledge, 212
Sociology, 242
Soil degradation, 266–67
Sophists, 76–79
Sophocles, 84, 85
Sorcerer (cave painting), 28
Spain, 110, 130, 133
Specialization, 287, 288, 298–99
  ancient, 39
  and Aristotle, 84
  and knowledge, 60, 142
  and printed literature, 138
  in science, 168
Speech. *See* Language
Sports. *See* Athletics
Statistical Society, 242
Statistics, 170, 229, 231, 232, 235–36, 241–42
Steam engine, 165, 194, 195
Steel, 195–96
Steele, Richard, 185
Stelluti, Francesco, 167
Stethoscope, 233
Stone bowls, 49
"Stories for the Middle Ranks of Society and Tales for the Common People" (More), 203
Strato of Lampsacus, 83
String, 17
Stylus, 50
Sulfur dioxide, 265
Sumeria and Sumerians, 43–50, 55, 310
Sun, 5
Sunday school, 206–7, 210
Surveying, 169
Suwa, Gen, 8
Sydenham, Thomas, 227
Syllabaries, 59–60
Sylvian fissure, 29
Syphilis, 246

Tacitus, 133
Taxonomy, 226–27
Technology, 176–77
Teeth, 38
Telescope, 165, 169
Temple, William, 215–16
Textbooks, 139, 140
Textile industry, 194, 219
Textual criticism, 135
Theater, 84–85
Theodoric of Freiburg, 118, 119
Theology, 114, 116–17
Thermometers, 232–33
Thinking
  alphabetic, 64–72
  arational, 293
  imaginative, 292–93
  sequential, 18–20
Thrasymachus, 79
Time, 215–16
Time keeping, 109
  See also Clocks
Tithe, 100
Tokens, 42–45
Tools and toolmaking

first primitive, 10–11
kits, 20
new approach to, 20–21
prehistoric, 15–17
techniques, 20–21, 26
See also specific tools, e.g., Carrier bags
Topsoil, 266–67
Torricelli, Evangelista, 162, 164
Trade, 46, 56, 171
law of, 188
The Tradesman's Calling (Steele), 185
Translation, 109–11
Trois Frères cave sanctuary, 28
Tropical forests, 269
Truth, 77, 81
according to Augustine, 95–97
Tuberculosis, 246
Tull, Jethro, 183
Typhoid, 247
Typhus, 236

Underclass, 210–11
Unemployment, 180, 205, 221
rural, 264
Unions, 216
United Nations, 304
United States, 173, 264, 265, 272, 286, 303
University of Wittenberg, 139
Urbanization, 201, 274
Ur-Engur (king of Ur), 54
Ur-Nammu (king of Nippur), 54
Uruk, 44, 47–48, 49

Vacuum, 162–64
Vacuum pump, 164
Venus figurines, 28–29
Vergil, 94
Vespucci, Amerigo, 147
"Village Politics" (More), 202–3
Virchow, Rudolf, 243–44

Wages, 191, 215
Waldseemüller, Martin, 147
Waste dumping, 265–66
Water
demand for, 261–62
pollution, 262–63, 266
unsanitary, 237
Watt, James, 195
Wealth of Nations (Smith), 172, 191
Web data-links, 293–96, 298, 301–2, 309
Weights and measures, 54
Welfare legislation, 180
Weston, Richard, 178
Wheat, 259–60
Wheel, 45
Wiliness, John, 170
Wilkinson, John, 195
William of Conches, 114
William of Moerbeke, 111
Wilson, C. T. R., 276
Witt, Jan de, 170
Women, 46, 305
Work. *See* Labor
World Health Organization, 265
Writing, 36, 205
alphabet development, 58–60
arrangements, 71
early forms of, 40–42, 50–52, 56–57
first example of, 42
left-to-right orthography, 71–72
in Mesopotamia, 286
See also Batons
Wunderlich, Karl, 234

Yahya (Mesopotamia), 49
Yeats, W. B., 292
Young, Arthur, 181

Zulu (people), 15

About the Authors

James Burke is an award-winning television host, author, and educator, best known for his extremely successful PBS series *Connections.* He has hosted many other PBS series, on topics such as the Greenhouse Effect, Renaissance painting and the human brain, as well as the recent *Connections*^{2}. He is the author of the best-selling companion books to *Connections* and other series, including *The Day the Universe Changed,* and is a regular columnist for *Scientific American.* He lectures around the world and divides his time between London and France.

Robert Ornstein is the author of more than twenty books, including *The Psychology of Consciousness, The Evolution of Consciousness, The Healing Brain* (written with David Sobel) and *New World, New Mind* (with Paul Ehrlich). He taught at Stanford and the University of California medical center for twenty years, and is the author of a leading psychology textbook.

About the Illustrator

Having trained as an engineer and taught physics, Ted Dewan became a professional illustrator and cartoonist in 1988. He has illustrated nonfiction children's books, including *Inside Dinosaurs and Other Prehistoric Creatures* and *Inside the Whale and Other Animals,* the latter of which won him the Mother Goose Award in 1992. *The Axemaker's Gift* is his third illustrated book with Robert Ornstein.